AF693209

# THE
# JAMES SPRUNT STUDIES
# IN HISTORY
# AND POLITICAL SCIENCE

*Published under the Direction of*
*the Departments of History and Political Science*
*of the University of North Carolina*

VOLUME 38

---

# AGRICULTURAL DEVELOPMENTS IN NORTH CAROLINA 1783-1860

*By*

CORNELIUS OLIVER CATHEY

CHAPEL HILL

THE UNIVERSITY OF NORTH CAROLINA PRESS

1956

*The Seeman Printery, Inc.*
*Durham, N. C.*

# PREFACE

My interest was drawn to ante-bellum agriculture as a promising field for research and study by remarks contained in his lectures on the Old South by Professor Fletcher M. Green, of The University of North Carolina. The particular topic selected is one in which it has been possible for me to feel a keen personal interest also, for, with the exception of the present generation, my family has wrested its livelihood from the soil of North Carolina for over two hundred years.

Although necessarily prolonged, because of the war, my study of this subject has been both interesting and revealing. Contrary to the opinion of one author that the South was "scarcely affected" by the improvements made in agriculture between 1783 and 1860,[1] I have found that nothing significant occurred in American agriculture during that time which was not reflected in parallel or corresponding developments in North Carolina. In the introduction of new farming implements alone, the period witnessed revolutionary changes concerning which North Carolina and North Carolinians had a significant part. To note the developments of this nature, rather than to record the history of the state's agriculture in this period, has been the principal aim of this study.

At numerous points in the development of this study and in the preparation of the manuscript, I have received valuable advice and direction from Professor Green. Also, the late Professor Albert R. Newsome, and Professors Hugh T. Lefler and J. Carlyle Sitterson of The University of North Carolina, have given assistance in many respects, particularly in helping the author to arrive at a proper appreciation of the North Carolina of the ante-bellum period. To these four, I am principally indebted for whatever of merit this study possesses. To myself alone, the imperfections are due.

Throughout the period spent on this study, sharing with me in impatience at the delays encountered, and proving a constant source of inspiration, my wife has encouraged and assisted me on every hand. In tribute to Beulah's constancy and devotion, I dedicate this volume to her.

CORNELIUS OLIVER CATHEY.

[1] Ernest Ludlow Bogart, *Economic History of the American People* (2nd ed.), 443.

# CONTENTS

# AGRICULTURAL DEVELOPMENTS IN NORTH CAROLINA

1783-1860

CHAPTER I

# THE LAND, THE PEOPLE, AND THE TRADITION

North Carolina lies between 34° and 36° 31′ north latitude and between 75° 30′ and 84° 15′ longitude west of Greenwich.[1] Roughly, the state is rectangular in shape and contains 52,250 square miles, of which over 3,000 are under water. It is 503 miles long and averages about 100 miles in width; at its widest point it has nearly 300 miles of frontage on the Atlantic Ocean.

The state contains three well defined geographic regions: the Atlantic Coastal Plain, which extends from the coast westward to the "fall line" of the rivers; the Piedmont, which lies between the fall line on the East and the foothills of the Appalachians on the West; and, the Mountains.[2] Differences of soil, elevation, rainfall, and length of growing season between these regions combine to give North Carolina an ability to produce, with some degree of success, practically every crop grown in the temperate zone.

With respect to the origins of the soils, North Carolina is about equally divided into two areas: the Coastal Plain, with its unconsolidated rocks and sedimentary soils; and, the Piedmont and Mountain regions, with their hard, crystalline rocks and residual soils. Several sub-divisions of these soil areas have been created by variations in soil types.[3]

[1] This location, being in the same belt of the globe as the citrus fruit and date producing regions of North Africa and the Near East, gave rise to the thought that similar crops could be produced in the state. Efforts were made in that direction during the colonial period, but without success. John Lawson, *History of North Carolina,* 44. The Lawson volume furnishes one of the best descriptions of the flora and fauna of early North Carolina, and, also, one of the first attempts at a survey of the state's agricultural potential. The best single volume coverage of North Carolina's history is to be found in Hugh Talmage Lefler and Albert Ray Newsome, *North Carolina: The History of a Southern State.* Cited hereinafter as *North Carolina.* See also, Hugh Talmage Lefler, *A Guide to the Study and Reading of North Carolina History.*

[2] Descriptions of the limits of these regions are contained in Ebenezer Emmons, *Geological Report of the Midland Counties,* 3-6; Robert Digges Wimberly Connor, *North Carolina: Rebuilding an Ancient Commonwealth,* I, 9-13, cited hereinafter as *Rebuilding an Ancient Commonwealth.*

[3] For descriptions of the soils of North Carolina, see Hugh Hammond Bennett, *The Soils and Agriculture of the Southern States,* 52-166, 176-192, *passim;* Milton Whitney, *Soil and Civilization: A Modern Concept of the Soil and the Historical Development of Agriculture,* 2, 34-41, 50-61, 69-71, hereinafter cited as *Soil and Civilization;* Samuel Huntington Hobbs, Jr., *North Carolina: Economic and Social,* 5-7. Appendices "A" and "B" of H. H. Bennett, *The Soils and Agriculture of the Southern States,* contain definitions of the various soils and discussions of classification of soil material by texture, grade, structure, color, and chemical analyses.

The richest soils are to be found in the alluvial bottom lands along the rivers of the Coastal Plain. The poorest are those of the Sand Hill and Pine Barren regions. Between these extremes, there are soils of every degree of productivity. Speaking of the state generally, Elisha Mitchell said in 1822: "Our grounds are neither so fertile that they will produce spontaneously what is necessary to the sustenance and comfort of our citizens, nor so sterile that we have reason to abandon them in despair."[4] Perhaps the best appraisal of the soils of the colony was given by Bishop August Spangenburg, who said: ". . . if a man would say it was half good—one fourth bad—one fourth 'middling' it would be correct. But all land in N. C. is so mixed, and no 600 acres can be taken up without some barrens."[5] Unquestionably, the principal deficiency in the soil has been, and is, a general lack of calcium compounds, which means the soil has an acid rather than an alkaline character. This factor varies in different areas and has had an important influence on the crops grown in each.

From the standpoint of agriculture, North Carolina is almost ideally served by nature in the amount, and, what is more important, the distribution of its rainfall. The average annual precipitation for the state is 50 to 60 inches, and the normal distribution is such that the crops get most when they need most. The normal year also offers a dry season for the harvest.[6]

North Carolina is well drained by its considerable network of rivers and their tributary streams. Most of these have their rise in the Piedmont and Mountains; and, with the exception of a few which eventually empty into the Gulf of Mexico, all empty into the Atlantic or the sounds along the coast. Principal among the rivers are the Catawba and Yadkin, which rise in the mountains and flow southeasterly into South Carolina where they are known as the Pedee and Wateree, respectively, and empty into the Atlantic; the Roanoke, its tributary the Dan, and the Chowan, which empty into Albemarle Sound; the Tar and Neuse, which empty into Pamlico Sound; the Waccamaw and Lumber, which flow into South Carolina, and empty into the Atlantic; and, the Cape Fear which empties into the Atlantic. The latter is the only major stream whose course lies wholly within the state and flows directly into the ocean. From the standpoint of inland transportation, the Cape Fear is the principal river.[7]

North Carolina enjoys a greater diversity of climate than any other

[4] Quoted in *Report on the Geology of North Carolina, Conducted Under the Direction of the Board of Agriculture,* Part I, 50-51.

[5] Quoted in *North Carolina Colonial Records,* V, 14. Cited hereinafter as *N. C. C. R.*

[6] S. H. Hobbs, Jr., *North Carolina: Economic and Social,* 11-12.

[7] James Dunwoody Brownson DeBow, *The Industrial Resources, etc., of the Southern and Western States: . . . . ,* II, 171. Cited hereinafter as *Industrial Resources.*

state on the Atlantic seaboard. The mean average temperature varies from 63° in the Southport area to 50° in the northwestern part of the state. There is no part of the state which has extremes of either cold or heat. The date of the first killing frost varies from thirty to forty days; and the length of the growing season varies from about 200 to about 240 days. The bulk of the area of the state has a growing season of something over 210 days.

> Taking all factors into consideration—extremes of heat and cold,—length of growing season, amount and regularity of rainfall, clear days, absence of storms, and so on—North Carolina averages up about as well as any area that can be presented. . . . While there are climatic variations, they are not extreme.[8]

The advantages these climatic factors offered for the diversification of agriculture were remarked upon by practically every ante-bellum writer on the state. James Dunwoody Brownson DeBow said:

> The industry of North Carolina is almost wholly agricultural. There is not a state in the Union more fortunate in its variety of staple productions. All kinds of grains that grow in the north are successfully cultivated here. The striking diversity of climate and soil . . . has its correspondence in a similar diversity of agricultural productions.[9]

"Alpha," of Edgecombe County, wrote that North Carolina's "position seems to be a choice medium between a northern and a southern climate—where the productions of both may flourish to advantage."[10]

Early accounts of North Carolina speak of the "stately timber" that abounded in and almost covered the entire area of the state like a "measureless forest, an ocean of trees."[11] Of the numerous varieties of both hard and soft woods which were to be found in this forest, the most important were wild cherry, dogwood, hickory, walnut, birch, ash, poplar, oak, cedar, chestnut, cypress, gum, and pine. Timber not only furnished the early settler with cheap shelter, warmth, and raw materials for furniture and implements, but also gave him the basis for an expanding industry, particularly in the pine forests for naval stores. These items, along with clapboards, staves, barrel headings, masts, shingles, and other lumber products constituted the principal export of the colony in the latter years of the colonial period and formed the basis of its economic prosperity.[12] The naval stores industry was carried on chiefly

[8] S. H. Hobbs, Jr., *North Carolina: Economic and Social,* 9-13.
[9] J. D. B. DeBow, *Industrial Resources,* II, 172.
[10] *Farmer's Journal,* II (October, 1853), 210.
[11] See for instance Johann David Schoepf, *Travels in the Confederation, 1783-1784,* II, 103. See also H. T. Lefler and A. R. Newsome, *North Carolina,* 22.
[12] Lewis Cecil Gray, *History of Agriculture in the Southern United States to 1860,* I, 45, 151-157, cited hereinafter as *History of Agriculture;* R. D. W. Connor, *Rebuilding an Ancient Commonwealth,* I, 34.

by farmers and planters as a side-line activity; it not only brought in sizable profits but also provided year around employment for slaves who might otherwise have been idle during part of the year.[13]

The forests of colonial North Carolina abounded in roots and herbs, which were relatively of greater significance then than now because of their universal employment in the treatment of diseases and "distempers" of all sorts.[14]

In addition to shelter, warmth, medicinal supplies, and income, the forests furnished the settler with food and clothing. Rabbits, fox, squirrels, raccoons, o'possums, beaver, otter, buffalo, elk, deer, bear, turkeys, ducks, pheasants, geese, wild pigeons, and wild bees were present in greater or less numbers.[15] Janet Schaw who visited in North Carolina at the very end of the colonial period, wrote that deer were "in great plenty" and gentlemen shot them "out of idleness"—nothing under a wild turkey being "worth a shot." The ducks were "the finest . . . that possibly can be met with," she continued, "and so plenty that when on wing sixteen or eighteen are killed at a shot."[16]

From all accounts the waters of the sounds, rivers, and lakes were equally as well stocked with fish. Unimpeded in that day by man made obstacles, fish migrated freely up and down the streams and provided inland, as well as coastal settlers, with food. The plenty of fish and game of the forest added to the provisions raised on the farm resulted in a very low living cost. The author of *American Husbandry* estimated "that one Indian, or dextrous negroe, will, with his gun and netts, get as much game and fish as five families can eat. . . ."[17] Undoubtedly, the ease with which a man might obtain food from the forests and

[13] "There was probably no colonial industry identified with frontier conditions in which slave labor could be more economically employed than in gathering of naval stores." Rosser Howard Taylor, *Slaveholding in North Carolina: An Economic View,* 12, cited hereinafter as *Slaveholding in North Carolina.* See also, H. T. Lefler and A. R. Newsome, *North Carolina,* 90-93.

[14] William Kenneth Boyd (ed.), *Some Eighteenth Century Tracts Concerning North Carolina,* 442-443, cited hereinafter as *Some Eighteenth Century Tracts.*

[15] *Ibid.,* 438; John Lawson, *History of North Carolina,* 44-45.

[16] Evangeline Walker Andrews and Charles McLean Andrews (eds.), *Journal of a Lady of Quality; Being a Narrative of a Journey From Scotland to the West Indies, North Carolina, and Portugal, in the Years 1774 to 1776,* 175-176. Cited hereinafter as *Journal of a Lady of Quality.*

[17] Harry J. Carman (ed.), *American Husbandry,* 303. Although purportedly written by "An American," this volume, published in London in 1775, has been variously credited to Arthur Young and to Doctor John Mitchell, an Englishman and member of the Royal Society who lived in Virginia for several years about the middle of the Eighteenth Century. *Ibid.*, xxxvi-xxxviii. Its editor has concluded that the author is, as yet, unidentified. Nevertheless, "Of all our colonial literature, *American Husbandry* is the most accurate and comprehensive account of the English colonies in America and gives by far the best description of their agricultural practices. The recommendations for the improvement of farming compare favorably with those of any modern textbook on the subject and are much superior in style and presentation to any other English or American agricultural books of the eighteenth century." *Ibid.,* xli. The writer concurs with this apparisal of *American Husbandry.*

streams of North Carolina—even a sumptuous living—had an enormous influence upon the character of agriculture developed in the colony prior to 1783.

Early in the history of the colony, objection was raised to the practice of granting large tracts of land to individuals, who had "not nigh people enough to manure and people the same," and the colonial assembly passed an act forbidding the taking up of tracts larger than 660 acres without the special permission of the lords proprietors.[18] In general, the proprietors and the king followed this policy with the result that North Carolina, more than the other southern colonies, came to be a colony of small landowners who lived on and cultivated their own lands.[19] By special dispensations, some people secured tremendously large holdings, but this was the exception rather than the rule.[20]

Because transportation was easier by water than by land, the tendency was to select sites along the rivers, preferably at points where there were natural "landings." This was particularly true of the settlements in the Coastal Plain. Unpatented lands were used as "commons," somewhat as in feudal times.[21]

Land was the principal form of wealth in colonial North Carolina and everything affecting ownership was of great concern to the people. Considerable difficulties arose as to the legality, registration, and failure or neglect to register titles; amounts, modes, time, and place of payment of quit rents; careless and inaccurate surveys; size and location of plots; and, the laws governing forfeitures and escheats.[22] With the exception of quit rents, which the settlers ceased paying upon the outbreak of the Revolution, the acquisition of political independence did not resolve these difficulties. The carelessness with which lands were surveyed, plots described, and boundaries marked proved a particularly prolific source of disputes and litigation after independence. Many surveys had to be redone several times with the result that much confusion arose.

The abundance of land and the ease with which it might be acquired are, perhaps, the two most significant factors influencing the character

[18] *N. C. C. R.*, I, 186.

[19] John Spencer Bassett, "Landholding in Colonial North Carolina," *The Law Quarterly Review*, XI (April, 1895), 154.

[20] The Moravian Church, for example, obtained nineteen deeds from Lord Granville for a total of 98,985 acres. John Henry Clewell, *History of Wachovia in North Carolina*, 12. "Orton" plantation, on the Cape Fear, the seat of "King" Roger Moore, contained more than 10,000 acres. Moore was called "King" because of his extensive holdings in land and slaves. James Sprunt, *Chronicles of the Cape Fear, 1660-1916* (2d. ed.), 40, 57-58. For references to other large holdings, see *North Carolina Minerva, and Raleigh Advertiser*, August 27, 1799, August 12, 1800; Edenton *State Gazette of North Carolina*, January 23, 1799.

[21] W. K. Boyd (ed.), *Some Eighteenth Century Tracts*, 444.

[22] Beverly Waugh Bond, Jr., *The Quit-Rent System in the American Colonies*, 114-117; Edwin Wexler Kennedy, *Quit-Rents and Currency in North Carolina, 1663-1776*, 13-15; *N. C. C. R.*, I, 170, 186-187; R. D. W. Connor, *Rebuilding an Ancient Commonwealth*, I, 97-98; J. S. Bassett, "Landholding in Colonial North Carolina," *loc. cit.*, 60.

of agriculture in colonial North Carolina. Generally, the holdings in the Piedmont were smaller than those in the Coastal Plain because of the topography of the land and the scarcity of large tracts suitable for plantation scale operations. In the Coastal Plain the smooth surface, the character of the soil, and the climate offered better prospects for and tended more towards large scale operations. This had definitely become apparent by 1783. These conditions, plus slave labor, meant the plantation system.[23]

Unlike those of her more fortunate neighbors, none of North Carolina's rivers are navigable by sea-going vessels, and at no time in her history has the state had an adequate harbor to serve as an entrepot for her trade and commerce. The rivers are generally broken by falls in their upper courses and are either too shallow or choked up by sand bars and debris at their mouths.[24] The direction of flow of North Carolina rivers, plus their very limited facilities for navigation, not only served to retard the spread of population into the interior but also the development of commercial agriculture.[25] It was in this connection that Professor Connor said: "Probably no state in the American Union has suffered more in its economic development on account of unfavorable geographic factors than North Carolina."[26] Unquestionably, the lack of markets or difficulty of access to them, which resulted from these geographic conditions, had as much influence in shaping the character of the state's agriculture, and in impeding efforts at its improvement as any other single factor. It was only in the last two decades of the period of this study, after an effective internal improvement program had been launched, that these obstacles began to be overcome. From that time forward, as the railroad came more into use, rivers became increasingly less important as routes of communication and commerce.

Practically all contemporary accounts of the roads and travel conditions in North Carolina, in the late colonial period, are filled with revelations of the authors' experiences of hardships, privations, delays, and dangers. Similar conditions affected travel in the other colonies, but they were particularly accentuated in North Carolina by reason of its great extent and sparcity of population. The roads were without system, being ill-defined, and, to a great extent, roads in name only. It was customary to blaze the trees at certain distances along the right-of-way to guide the traveler on the way; nevertheless, travelers frequently lost their way. One of Elkanah Watson's companions suffered such an

[23] L. C. Gray, *History of Agriculture,* II, 93; R. D. W. Connor, *Rebuilding an Ancient Commonwealth,* I, 13.

[24] John Spencer Bassett, "The Influence of Coast Line and Rivers on North Carolina," *Annual Report of the American Historical Association* (1908), I, 58-61. Bassett described the Albemarle and Pamlico Sound regions as being the most isolated part of the Atlantic coast between Savannah and Penobscot.

[25] L. C. Gray, *History of Agriculture,* I, 118.

[26] R. D. W. Connor, *Rebuilding an Ancient Commonwealth,* I, 17.

experience on a trip through the state in 1778, and, as a result, spent the night, unknowingly, within sight of Wilmington "fighting wolves with pitch-pine flaming [illegible]s."[27] Describing his trip from Newbern to Wilmington, which was over the main route through the state, Watson said: "I journeyed the entire day alone, through a wilderness of pines, over a flat, sandy country, with scarcely an inhabitant to be seen. . . . As night closed upon me, I was totally bewildered, and without a vestige of a road to guide me."[28]

There were few bridges in North Carolina in 1783. In the low country, however, ferries had been established at the main crossings of the rivers and sounds. These were operated on a toll basis. In the back country, river crossings were made at fords, most of which had been located by that time.[29]

For the farmer living in isolated regions, the profits involved in trade were so small that it was scarcely worth his time to produce more than enough to meet his own needs, and those of his immediate community. Such a course, however, not only tended to accentuate the isolation of these people, but also tended to develop in them narrow, and provincial attitudes as they lost the stimulation of contacts with the outside world. As the horizon of the frontier closed in about them, the people fell more and more into ignorance and superstition. The Moravians, who settled in present day Forsyth County, more than any other group, recognized the dangers of isolation and longed for better transportation, communication, and trading facilities.[30]

North Carolina made very little progress in the colonial period in the development of what might be called urban centers. Her most important villages were Halifax, Edenton, Newbern, and Wilmington along the coast, and Cross Creek (later Fayetteville), Hillsborough, Salisbury, and Charlotte in the back country. John Skinner, Marshal of the North Carolina District for the taking of the census in 1790, said he was "satisfied that not one town in North Carolina contains more than 2000 inhabitants."[31] This, according to present day standards, means that North Carolina, in 1790, was 100 per cent rural. This absence of towns contributed towards the retardation of development in the state of commercial, educational, and cultural institutions which are

[27] Winslow C. Watson (ed.), *Men and Times of the Revolution; or, Memoirs of Elkanah Watson, Including Journals of Travels in Europe and America, From 1777 to 1842, with his Correspondence with Public Men and Reminiscenses and Incidents of the Revolution*, 57. Cited hereinafter as *Memoirs of Elkanah Watson.*

[28] *Ibid.*, 40.

[29] W. C. Watson (ed.), *Memoirs of Elkanah Watson*, 38-40, 253, 256.

[30] Adelaide L. Fries (ed.), *Records of the Moravians in North Carolina*, II, 539-540. Cited hereinafter as *Records of the Moravians*. The German settler generally exhibited less willingness to do without such facilities. According to a contemporary observer: "They have no idea of sitting down in a wilderness, and growing wild in it." Harriett Martineau, *Society in America*, II, 40.

[31] *U. S. Census of 1790*, 49.

vital to the progress of a people. Throughout the period of this study, the state retained its almost complete rural character, and the principal interest of its people continued to be overwhelmingly agricultural.

According to the census of 1790, the total population of North Carolina was 393,751, of whom, 288,205 were whites, 100,570 were slaves, and 4,976 were in the category of "other free persons." Of the total population, 159,756 lived in the three western districts of the state, which comprised generally the area west of the fall line; and, of this number, 24,337, or fifteen per cent, were slaves. A total of 233,995 lived in the five districts comprising the area east of the fall line, or the Coastal Plain; and, of this number, 76,234, or thirty-two per cent, were slaves.[32]

Based upon indications contained in the names of heads of white families, which were taken in the census of 1790, 83.1 per cent were of English origin, 11.2 per cent were Scotch, 2.8 per cent were German, 2.3 per cent were Irish, 0.3 per cent were French, 0.2 per cent were Dutch, and all other nationalities were 0.1 per cent.[33]

At the end of the colonial period, the areas of heaviest settlement were the eastern half of the Coastal Plain and the Piedmont. Between the two areas of heaviest settlement lay the region of pine barrens which supported a very sparse population. A traveler through that area in 1769 wrote that he had traversed thirty or forty miles of the country without seeing a single house.[34] West of the Blue Ridge, a small number of the more adventurous pioneers had reached the "western waters" by 1783, and had staked out claims to lands in what is now Tennessee.[35]

Of the major racial elements comprising the population, the English had settled in the colony first and were principally located in the Coastal Plain.[36] The Scotch-Irish began to arrive in large numbers in the Piedmont about 1730, and immediately set themselves up on relatively small holdings as farmers although there were in their number capable weavers, joiners, coopers, wheel-rights, wagon-makers, tailors, blacksmiths, hatters, merchants, laborers, wine-makers, rope-makers, and fullers.[37] The Scotch-Irish came to the colony chiefly from Pennsylvania Virginia, and New Jersey. Following the same general route as the Scotch-Irish, a large number of Germans entered the colony after

[32] *Ibid.*

[33] L. C. Gray, *History of Agriculture,* I, 93; Guion Griffis Johnson, *Ante-Bellum North Carolina: A Social History,* 9, cited hereinafter as *Ante-Bellum North Carolina.*

[34] L. C. Gray, *History of Agriculture,* I, 119.

[35] Ulrich Bonnell Phillips (ed.), *Plantation and Frontier Documents, 1649-1863,* I, 84.

[36] For an account of the origin, characteristics, and contributions of the English element to the development of North Carolina, see R. D. W. Connor, "Race Elements in the White Population of North Carolina," *North Carolina State Normal and Industrial College Historical Publications,* No. 1, 20-43. Cited hereinafter as "Race Elements."

[37] R. D. W. Connor, "Race Elements," *loc. cit.,* 69-90.

1740 and settled in the region now represented by Orange, Rowan, Guilford, Burke, Lincoln, Randolph, Iredell, Stokes, Cabarrus, Davidson, Stanly, Catawba, Alamance, and Forsyth counties.[38] By habit and training, the German was thrifty, industrious, and frugal. Unaccustomed to slavery, he was more inclined to rely on his own labor. He soon became the best farmer in the colony.[39] A traveler described the Moravians in 1783 as "distinguished above the other inhabitants for their industry and diligence in agriculture and the crafts. A service which is recognized by most of their fellow-citizens but tempts very few to imitation."[40] The Highland-Scotch entered the colony after 1729 and settled in the region now comprised within Anson, Bladen, Cumberland, Harnett, Moore, Richmond, Robeson, Sampson, Scotland, and Hoke counties. Most of them had been farmers in Scotland, and they set themselves up as farmers in North Carolina.[41]

Because North Carolina was slow in developing a strictly commercial type agriculture, and because the Negro slave proved superior in the long run as a farm laborer, there never was a very considerable number of white servants in the colony. Many former indentured servants, however, were among the settlers who came down from Virginia.[42]

Slaves were introduced into the colony at an early date. Their numbers grew slowly, however, until the last thirty years of the colonial period, when a tremendous increase was made.[43] Most of the settlers were men of little capital, and, having inferior commercial facilities "[they] resorted to small-scale farming in which slave labor was an important, but by no means a controlling factor."[44] In the last years of the colonial period, a trend toward slave-holding became noticeable in the inland counties, particularly Granville, and those along the Virginia border which were producing larger quantities of tobacco.[45] Also, in the latter part of that period, some opposition to slavery arose, and, in 1774, the Provincial Congress resolved not to import any slaves or to buy any brought into the colony "from any part of the world."[46] This resolution was enforced for a short time; but the slave trade was resumed after the Revolutionary War.[47] By 1783, the institution of

[38] *Ibid.*, 91-111.

[39] L. C. Gray, *History of Agriculture*, I, 122.

[40] J. D. Schoepf, *Travels in the Confederation, 1783-1784*, II, 154.

[41] For origin, characteristics, and contributions of this group, see R. D. W. Connor, "Race Elements," *loc. cit.*, 11.

[42] John Spencer Bassett, *Slavery and Servitude in the Colony of North Carolina*, 15-16, cited hereinafter as *Slavery and Servitude;* L. C. Gray, *History of Agriculture*, I, 349.

[43] J. S. Bassett, *Slavery and Servitude*, 15; L. C. Gray, *History of Agriculture*, I, 355, II, 1025.

[44] R. H. Taylor, *Slaveholding in North Carolina*, 10.

[45] L. C. Gray, *History of Agriculture*, I, 355, II, 1025.

[46] *N. C. C. R.*, IX, 1046.

[47] R. H. Taylor, *Slaveholding in North Carolina*, 21-23.

slavery had reached "a stage of permanency and became a controlling factor in the social and economic life" of North Carolina.[48]

In order of social standing, the white population in 1783 contained the following classes: gentry, small farmer, and indentured servants. The gentry was the most influential, wealthy, and cultured of these three groups. Included in this category were the large planters, professional men, and higher public officials. They prided themselves in the titles of "Gentlemen," "Esquire," and "Planter." The planters made up the largest element of this class. They were to be found principally occupying the best sites along the Cape Fear, Tar, and Roanoke rivers in the East, but here and there along the river bottoms in the Piedmont were units approaching the plantation scale. In North Carolina, the gentry class probably represented a smaller percentage of the total white population than in any other of the southern states.[49] Possessing more education, capital, leisure time, and contacts with the outside world, the gentry class not only gave tone and direction to society and politics in North Carolina, but also contributed much more toward the development of agriculture than their numbers would indicate. They had more incentive to make such contribution than the small farmers. With them, as we shall see, it became to a great extent a case of improving their agriculture or moving, and moving for them was a much more complex problem than with the farmer.

The small farmer and artisan class, which was by far the largest class, came next in the social scale. There were in this class all of the farmers, whether they owned slaves or not, who were principally engaged in self-sufficing or subsistence farming.[50] They prided themselves in the title of "Farmer," "Husbandman," and "Yeoman," and, though generally inarticulate, they were the authors of most of the articles that appeared later in farm periodicals signed "Clod-hopper," "Experience," "Hard Times," etc., just as surely as those articles signed "Alpha," "Cincinnatus," and "Agricola" may be ascribed to the usually classic-minded gentry. In the artisan class, too, were the workers in the naval stores industry, small merchants, blacksmiths, overseers, gunsmiths, tutors, drovers, and tavernkeepers. From the lower level of this class came, at a later date, that ignorant, mean, despicable, and diseased white man referred to by the Negroes as "poor white trash."[51]

[48] *Ibid.*, 10.

[49] R. D. W. Connor, *Rebuilding an Ancient Commonwealth,* I, 164-165; L. C. Gray, *History of Agriculture,* I, 493; Ulrich Bonnell Phillips, *Life and Labor in the Old South,* 94-95. For good descriptions of the various social classes in the population of the antebellum South, see D. R. Hundley, *Social Relations in Our Southern States.*

[50] R. D. W. Connor, *Rebuilding an Ancient Commonwealth,* I, 165-166. D. R. Hundley in his *Social Relations in Our Southern States* included in this class all owners of less than fifty slaves. *Ibid.,* 84-85.

[51] W. K. Boyd (ed.), *Some Eighteenth Century Tracts,* 449. For an able discussion of "poor white trash," see D. R. Hundley, *Social Relations in Our Southern States,* 250-283.

One writer has said that during the colonial period the American farmer made three major accomplishments. Firstly, he learned from the Indian all that the Indian knew about agriculture, particularly his methods of clearing and bringing the land into cultivation. Secondly, the colonial farmer had experimented with numerous crops, some with which he had been familiar in the Old World, others that were native to the New World, and still others, chiefly semi-tropical crops, which it was thought might be profitably grown in the land. As a result of that experimentation he had obtained a great mass of information as to the limitations imposed by the soil, climate, and labor factors. Lastly, though the system of agriculture adopted by the colonial farmer was the most profitable for him, it led to such great waste of the soil and forests that it amounted to a butchery of the land.[52] This appraisal might very well have been made for the colonial North Carolina farmer alone; for, by 1783, he had experimented with practically every crop grown in the state at the present time and several that are not now grown, and, he had adopted practices which were impoverishing his soil and making him a wanderer in his own land.[53]

Land was plentiful and easily obtained in colonial North Carolina. The settlers, therefore, chose to adopt the economy of extensive agriculture wherein a minimum of labor and capital were employed per given unit of land. Labor and capital were the scarce factors. The soil was relatively rich and did not require careful or scientific tillage to produce substantial returns. When its fertility became wasted, the farmer simply abandoned the old and brought new acres under cultivation. The system was one wherein fields were rotated rather than crops. "This is the system," prophesied a contemporary, "of all the provinces from New England to Florida; and it is a system from which they will all by and by feel the most monstrous inconvenience."[54] Land bore the major portion of the burden of production and was used wastefully. "This," according to one critic, "is the key to an understanding of the agricultural history of the colonial period. . . ."[55]

Clearing new ground, therefore, became an annual chore for the average farmer or planter in colonial North Carolina. Although methods varied, the usual implements were the axe, the mattock, and fire. Most of the work of clearing new grounds was done in the winter, and, to a considerable extent, planters depended upon it to provide employment for the slaves in the off seasons. Saw mills were used during the period,

[52] Albert Hart Sanford, *The Story of Agriculture in the United States*, 55-56.

[53] In 1783, about as much land had been abandoned in North Carolina as no longer sufficiently productive as was then being cultivated. As a contemporary explained, this condition resulted from "the strange management they have of roaming from piece to piece and touching none without ruining and exhausting it." H. J. Carman (ed.), *American Husbandry*, 258-259.

[54] *Ibid.*, 317.

[55] Ernest Ludlow Bogart, *Economic History of the American People* (2d. ed.), 39, 47-48.

but most of the timber not needed on the farm was burned where it fell. Tremendous waste of forest growth resulted from the clearing process, and, also, from the practice of burning over the woodlands in the late winter so that a fresher growth of grass might be provided the cattle in the spring.[56]

Subsistence crops were of first importance to the colonial farmer, and, of these, Indian maize, or corn, came nearest to fulfilling the food requirements of both man and beast. John Lawson declared corn "the most useful grain in the world" and set forth strong substantiation for his opinion.[57] Unquestionably, it was the most valuable crop grown in colonial North Carolina, and "every proprietor of ever so small a piece of land [raised] some Indian corn. . . ."[58] Fortunately, it could be and was grown in every section of the state.

Practice in the cultivation of corn varied with individuals, sections, and soils, but in the main the work was performed by hand. In the older settled sections, by 1783, use of plows and draft animals had become more common. Like the Indian, the colonial farmer planted his corn in hills, which were spaced at intervals of about three feet with little regard for alignment into rows. These hills were dug up with hoes, and, if available, some farmers put a fish in each hill as a fertilizer.[59] From six to ten grains were planted in each hill and later thinned down to two or three stalks.[60] During the growing season, suckers were pulled from the corn and the hoe used to draw dirt up around the stalk. The Indian practice of planting beans, peas, melons, and squash with the corn was widely used also. In succeeding years, the practice was to dig up the same hills and to plant corn in them again. When fertility was exhausted, more lands were cleared and brought into cultivation. Practically no effort was expended to prevent soil exhaustion or to restore fertility to worn-out lands.[61]

The colonial farmer stored his corn harvest in a crib which was of Indian design. This building had larger inside measurements at the top

[56] L. C. Gray, *History of Agriculture*, I, 198. Of the practice of burning over the grazing lands, Janet Schaw wrote in 1774: "This morning a fine wood was set on fire just by us, and tho' I was informed of the reason and necessity of it, yet I could not look at it without horror; before it could be extinguished twenty thousand trees at least must have been burnt." E. W. Andrews and C. M. Andrews (eds.), *Journal of a Lady of Quality*, 149.

[57] John Lawson, *History of North Carolina*, 42-43.

[58] E. W. Andrews and C. M. Andrews (eds.), *Journal of a Lady of Quality*, 280-281.

[59] Louis Bernard Schmidt and Earle Dudley Ross (eds.), *Readings in the Economic History of American Agriculture*, 50-51.

[60] L. C. Gray, *History of Agriculture*, II, 814. The Indian understood and practiced the art of testing his seed for germination possibilities. He also used preparatory germination by steeping the seed corn in warm water. These practices were taken up by the white man. L. B. Schmidt and E. D. Ross (eds.), *Readings in the Economic History of American Agriculture*, 50-51.

[61] Lyman Carrier, *The Beginnings of Agriculture in America* (1st. ed.), 198-199; E. L. Bogart, *Economic History of the American People* (2d. ed.), 46.

than at the bottom, was elevated above the ground upon stilts or pillars, and offered reasonable safeguard against the weather, rats, and other foragers.[62] While in the husk, the matured corn withstood the weather very well, and could be harvested almost at the farmer's leisure. By 1783, the more provident farmers had begun to gather the fodder and tops for use as forage for their livestock. Others simply turned their livestock into the corn fields to forage for themselves, after the ears had been pulled from the stalks and removed.[63]

Although other cereals, wheat, rye, oats, millet, buckwheat, and barley were grown, corn completely overshadowed them all in importance to the North Carolina farmer in 1783. It could be grown on a wider variety of soil types. It was better adapted to the pioneer's system of culture. It was a hardier plant. It was of more benefit to both man and animal. And it required no elaborate or expensive milling to prepare it for use.

The soils and climate of the Coastal Plain were not well adapted to the growth of wheat, oats, rye, buckwheat, millet, and barley and not much was grown there; however, in the Piedmont, where conditions were favorable, culture of these grains was becoming increasingly important at the end of the colonial period. The Diary of the Salem Congregation for 1775 carries this entry: "Our neighbors are steadily increasing the amount of wheat raised. . . ."[64] There were, however, many indications that North Carolina had not yet reached the point where she produced her own requirements of grain.[65] The colonial's preference for corn served to reduce the acreage devoted to the smaller grains.[66]

In the newer settled regions, the smaller grains were still being planted in hills and cultivated entirely with the hoe in 1783. The seed was prepared for planting by soaking it in a strong salt-lye solution, and then mixing it with shell lime in an effort to get rid of cheat, darnel, and other weeds, and as a safeguard against fungus diseases.[67] The harvesting was done almost entirely with the hand sickle, and threshing was accomplished either with the flail, or the grain was trodden out by livestock.[68] Here and there in the state, bolting mills had been located for converting the grain into flour.[69]

[62] L. B. Schmidt and E. D. Ross (eds.), *Readings in the Economic History of American Agriculture,* 50-51.

[63] L. C. Gray, *History of Agriculture,* I, 173-174.

[64] Adelaide L. Fries (ed.), *Records of the Moravians,* II, 884.

[65] *Ibid.,* II, 730, 734, 816; *N. C. C. R.,* V, xlii; E. W. Andrews and C. M. Andrews (eds.), *Journal of a Lady of Quality,* 280-281.

[66] John Lawson, *History of North Carolina,* 42.

[67] L. C. Gray, *History of Agriculture,* I, 170-171. Darnel is an annual grass which is often found growing as a weed in grain fields.

[68] *Ibid.,* I, 161, 170-171.

[69] Adelaide L. Fries (ed.), *Records of the Moravians,* II, 884; *N. C. C. R.,* V, xlii.

The soils of North Carolina were suited to both the upland and lowland varieties of rice and some of each was grown. Rice growing, however, never attained the proportions in the state that it did in South Carolina, where it had become "the grand staple" during the colonial period. North Carolina's location coincided with the northern limits of the climatic region suited to the culture of rice, and, therefore, rice growing was more hazardous.[70]

In 1783, the culture of rice in North Carolina was fairly well confined to the maritime parts of the state, particularly to the region drained by the lower Cape Fear River. There, sites were selected along the streams which emptied into the ocean, above the point where the fresh water met the salt water, but within the area where the level of the water was affected by tidal action. Such a location permitted, with the aid of dikes, flood gates, canals, and ditches, an efficient control of the necessary flooding and draining of the rice fields during the growing season. Proper maintenance of these controls, however, necessitated not only a constant care on the part of the planter, but also the employment of a tremendous amount of labor under very trying, and always dangerous, working conditions. The major portion of the labor in the rice fields fell to the lot of slaves. Practically all of this labor was performed with hand implements, principally the hoe, mattock, and shovel, animal drawn implements not coming into use in the rice fields until a later date. A high quality of rice was produced in the state, and a goodly portion of it sold to South Carolina planters for use as seed. From South Carolina had come the original impulse towards culture of the crop, and most of the techniques employed were modeled on those in use in that state.[71]

Production of vegetables did not become a very important activity on the North Carolina farm prior to 1783. Many of the people still lived on the frontier and were not permanently settled. There were practically no markets for surpluses, and knowledge and facilities for preserving vegetables were very limited. Then, too, meat made up a proportionately larger share of the diet, and meat could be had in great variety and plenty for practically nothing. These factors made vegetable culture a very minor farm activity.[72] The colonial period was not devoid of accomplishments in this field, however. Many varieties of European vegetables had been introduced and native varieties adopted.[73] John

[70] H. J. Carman (ed.), *American Husbandry*, 244.

[71] H. J. Carman (ed.), *American Husbandry*, 244, 275-281; W. Neil Franklin, "Agriculture in Colonial North Carolina," *North Carolina Historical Review*, III (October, 1926), 556-558.

[72] E. W. Andrews and C. M. Andrews (eds.), *Journal of a Lady of Quality*, 174-179; W. K. Boyd (ed.), *Some Eighteenth Century Tracts*, 442-443; John Lawson, *History of North Carolina*, 43-44.

[73] John Lawson, *History of North Carolina*, 43; E. W. Andrews and C. M. Andrews (eds.), *Journal of a Lady of Quality*, 174-175.

Lawson described the gardeners in North Carolina as "negligent and unskilful."[74]

Of the vegetable crops, the sweet potato proved of greatest value to the early settler and to the people of the state during the period of this study. It was used as a food for man and livestock and relished by both. Five principal varieties were grown: the "Spanish," which was coarse and fibrous and best suited for animal food; the "Carolina," which was of little better quality; the "Brimstone," of middling quality and so-called because of its red skin; the "Purple," also named for its color; and, the "Bermuda," which was a round white potato, and, with the "Purple," was best suited for human consumption.[75]

Other vegetables grown in the state were carrots, leeks, parsnips, turnips, "love apples" or tomatoes (used for ornamentation only), Irish potatoes, artichokes, radishes, beets, onions, shallots, chives, pumpkins, squashes, cucumbers, melons, lettuce, cabbage, spinach, fennel, purslane, samphire, rhubarb, rocket, cress, parsley, asparagus, watermelons, cantaloupes, coleworts, and numerous varieties of peas and beans.[76]

The fine adaptability of the soils and climate of North Carolina for the production of peas and beans was recognized early, and their production attained some commercial importance in the colonial period. They were not in the category of "enumerated articles," which had to be sent direct to England, but were in the general category of "provisions," which the colony exported principally to the West Indies.[77] Elkanah Watson visited a planter near Halifax, in 1778, who had been producing around 4,000 bushels of peas annually.[78] The value of these legumes, particularly the old-field or common pea, for use in rotation schemes to maintain soil fertility, was not known in the colonial period, and none was planted for that purpose. Peas were planted almost universally with the corn, and were widely used as a food by man and livestock.

By 1783 all of the common English fruit trees had been introduced into North Carolina, and, where soil and climate were suitable, were found to thrive.[79] Elkanah Watson, in 1778, spoke of the land between Wilmington and Tarborough as abounding in "luxuriant peach or-

[74] John Lawson, *History of North Carolina,* 43-44.

[75] L. C. Gray, *History of Agriculture,* I, 4, 193.

[76] John Lawson, *History of North Carolina,* 43-44; L. C. Gray, *History of Agriculture,* I, 4-5. Some of these vegetables were later abandoned and are not now being grown in the state. Of these, fennel and samphire were European vegetables of the carrot family. Fennel, however, was grown principally for the aromatic flavor of its seeds. Rocket is a variety of colewort or cabbage. Purslane is a plant with succulent leaves and is used in salads.

[77] John Lawson, *History of North Carolina,* 43; Charles Christopher Crittenden, *Commerce of North Carolina, 1763-1789,* 39, 59.

[78] W. C. Watson (ed.), *Memoirs of Elkanah Watson,* 58.

[79] John Lawson, *History of North Carolina,* 63; E. W. Andrews and C. M. Andrews (eds.), *Journal of a Lady of Quality,* 163; W. N. Franklin, "Agriculture in Colonial North Carolina," *loc. cit.,* 562.

chards."[80] The author of *American Husbandry* said: "Fruit in none of the colonies is in greater plenty [than in North Carolina], or finer flavour; they have every sort that has been hitherto mentioned in this work; peaches, as in the central colonies, are so plentiful, that the major part of the crop goes to the hogs."[81] There was no market for surpluses of fruit, and no means of preserving it except by drying. By 1783, quantities of apples and peaches were being used in the making of brandies. Generally, the orchards were poorly kept and hogs and cattle foraged in them at will.

Other fruits grown in colonial North Carolina were figs, cherries, apricots, pears, plums, pecans, quinces, damsons, nectarines, and grapes. Some were native to the region, and others, like the fig and the peach, were introduced by the settlers. The forests abounded in butternuts, hazelnuts, chinquapins, chestnuts, walnuts, and hickory nuts.[82]

A great variety of wild berries grew in North Carolina and made up an important part of the farmers harvest. Of these, the most important were dewberries, blackberries, raspberries, huckleberries, strawberries, and mulberries. This last gave rise to the hope that a silk industry might be profitably established in the state.[83]

Except for the cereals previously mentioned, there was no cultivation of grasses in North Carolina in the colonial period for either forage or soil building purposes. The great abundance of "commons" lands, together with the practice of the "open range" combined to make the culture of artificial grasses of little consequence. Then, too, the soils of the Coastal Plain, where the bulk of the population was located, were not well suited to the culture of grasses without special fertilization.[84]

The soil and climate of North Carolina were adapted to the culture of hemp and flax, and several efforts were made in the colonial period to encourage their culture but with little success. This development had not passed the experimental stage when it was cut short by the Revolution.[85]

Likewise, efforts were made to develop the indigo industry. However, the culture of this crop was delimited by its peculiar climatic requirements. Indigo suffers particularly from droughts, both during the growing season and in the period when the plant is being processed for the market. Its successful culture, too, was hampered by plagues of

[80] W. C. Watson (ed.), *Memoirs of Elkanah Watson*, 58.

[81] H. J. Carman (ed.), *American Husbandry*, 243.

[82] W. K. Boyd (ed.), *Some Eighteenth Century Tracts*, 442-443; John Lawson, *History of North Carolina*, 63-65; L. C. Gray, *History of Agriculture*, I, 5.

[83] E. W. Andrews and C. M. Andrews (eds.), *Journal of a Lady of Quality*, 152; H. J. Carman (ed.), *American Husbandry*, 252-253.

[84] L. C. Gray, *History of Agriculture*, I, 177-179.

[85] *Ibid.*, I, 180-181; W. N. Franklin, "Agriculture in Colonial North Carolina," *loc. cit.*, 558; Newbern *North Carolina Magazine; or, Universal Intelligencer*, July 13, 1764.

grass-hoppers and caterpillars, which the plant seems to attract. North Carolina, also, lay on the outer fringe of the climatic region in which indigo may be successfully grown. Therefore, in spite of the bounty offered for the production of indigo its culture did not become widespread or commercially important in the colony; and, when the Revolution brought an end to the bounty system, indigo production ceased in North Carolina.[86]

In the interest of her own woolen and linen industries, England discouraged the production of cotton in her colonies. There was no objection, however, to its being grown for home consumption, and its culture was widespread in North Carolina, if on a small scale, prior to 1775.[87] Janet Schaw observed that cotton was "troublesome" to grow, gather, and separate from the seed, but, if managed properly, "it would be an Article of great consequence."[88] The author of *American Husbandry* wrote that "Cotton does very well, and the sort is so excellent that it is much to be wished they had made a greater progress in it."[89] The British blockade during the Revolution stimulated cotton growing in the state. However, the full development of this crop awaited the evolution of new techniques of production, processing, and manufacturing which came in the subsequent period.

North Carolina turned early to the culture of tobacco, and, throughout the colonial period, it was her most important commercial crop. The North Carolina growers drew heavily on Virginia experience with the crop, and, to a great extent, the tobacco growing region of the colony may be considered as but an extension of the Virginia region. Most of the North Carolina crop was marketed in Virginia markets, since the colony had no suitable port of her own, and the tobacco area was contiguous to Virginia. In North Carolina, however, culture of tobacco never became the all-absorbing interest to producers that it did in colonial Virginia and Maryland. North Carolina lay too remote from markets. Her transportation handicaps added too much to the costs and hazards of production. And, being a competing crop with Virginia's "grand staple," North Carolina's tobacco was discriminated against in the Virginia markets.[90]

The tobacco planters, like the planters of other commercial crops,

[86] *N. C. C. R.*, V, xli; H. J. Carman (ed.), *American Husbandry*, 281-287, 305; E. L. Bogart, *Economic History of the American People* (2d. ed.), 58; L. C. Gray, *History of Agriculture*, I, 295. For an account of the methods of production and processing of indigo, see Julian Alvin Carroll Chandler, and others (eds.), *The South in the Building of the Nation: A History of the Southern States Designed to Record the South's Part in the Making of the American Nation. . . .*, V, 178-183, cited hereinafter as *The South in the Building of the Nation.*

[87] L. C. Gray, *History of Agriculture*, I, 182-183.

[88] E. W. Andrews and C. M. Andrews (eds.), *Journal of a Lady of Quality*, 195.

[89] H. J. Carman (ed.), *American Husbandry*, 236-237.

[90] *N. C. C. R.*, V, 1-2.

sought out sites along the navigable streams which would serve them as highways to market for their product. This practice led to the rapid engrossment of lands affording such facilities, and tended to relegate the self-sufficing farmer to the less desirable locations. North Carolina was less fortunate in this respect than Virginia or Maryland in that she did not have many such locations.

Tobacco was found to grow best in freshly cleared woodlands which had accumulated through processes of vegetable decay a surface stratum of rich dark mould. Tobacco proved to be a quick exhauster of the soil in which it was grown.[91] This was due principally to two causes: firstly, in the harvest, the entire plant was cut and removed from the field; and, secondly, the cultivation was very shallow, merely loosening the surface soil, which made it readily subject to erosion by rain. Practically no effort was made to prevent this soil wastage prior to 1783.

A regular feature, therefore, of the colonial tobacco plantation's annual routine was the clearing of wood lands for use as new planting sites. "Such land will, after clearing," wrote the author of *American Husbandry,* ". . . prove more profitable to the planter than the power of dung can do on worse lands: This makes the tobacco planters more solicitous for new land than any other people in America, they wanting it much more."[92] It was thought that planters should have fifty acres of land for each working hand, for "with less than this they will find themselves distressed for want of room."[93] Lacking new lands into which to expand, the tobacco planter sold his old land to a new settler for use as corn or wheat lands, and moved with his Negroes, cattle, and tools toward the Piedmont. In North Carolina, in 1783, there was a strong tendency in that direction among tobacco growers.[94]

North Carolina had its greatest agricultural distinction in the colonial period as a producer of livestock, particularly cattle and swine. The author of *American Husbandry* observed that it was "not an uncommon thing to see one man the master of from 300 to 1,200 and even 2,000 cows, bulls, oxen, and young cattle; hogs also in prodigious numbers."[95] None of the livestock was native to America. All had been introduced into the colonies by Europeans. Among those cattle introduced were the Spanish blacks, the Devonshire reds, the Dutch black and whites, and

[91] "There is no plant in the world," wrote the author of *American Husbandry,* "that requires richer land, or more manure than tobacco; it will grow on poorer fields, but not to yield crops that are sufficiently profitable to pay the expenses of negroes, & c." H. J. Carman (ed.), *American Husbandry,* 164-165. See also, Avery Odell Craven, *Soil Exhaustion as a Factor in the Agricultural History of Virginia, 1606-1860,* 32. Cited hereinafter as *Soil Exhaustion.*

[92] H. J. Carman (ed.), *American Husbandry,* 164-165.

[93] *Ibid.,* 165.

[94] H. J. Carman (ed.), *American Husbandry,* 162-163, 165; L. C. Gray, *History of Agriculture,* I, 85.

[95] H. J. Carman (ed.), *American Husbandry,* 251.

the Danish yellows, with some additions from Ireland, Wales, and Sweden. By 1783, no distinctive American breed had been developed.[96] In North Carolina, because of the conditions under which the industry was conducted, the tendency was for livestock to deteriorate in breed. Generally they were suffered to shift for themselves, both in summer and winter, and very little if any attention was given to their shelter, feeding, or breeding. As a result of several generations of such neglect, livestock became stunted and scrawny, having the appearance of "runts."

In 1783, the settlement of the state was sparse, and vast tracts of land affording abundant mast, roots, and wild grasses served the people as a great "commons" upon which their cattle, sheep, horses, and swine foraged. There were few fences and quite often the only indication of a property line was a "notch" on a tree. Each owner chose a peculiar brand and mark to identify his property, and had them recorded at the county seat.[97] Periodically, the livestock were rounded up into pens or corrals where ownership was determined and the branding done.[98] A traveler in the state, in 1783, wrote that many livestock roamed unmarked and "wild" in the forests, being no man's property except him on whose land they were found. "But in certain parts," he continued, "there is a 'woods-right' so-called—according to which every plantation has a fixed share of all wild herds thereabouts; and this right, like any other property, can be transferred or sold at will."[99] Farmers, who wished to make their crops secure against foraging and trampling by livestock, fenced their fields. The livestock owner was under no compulsion to fence the stock in.

Some owners sought to attach their livestock to the home place by putting out salt for them, by penning their young during the day, and by calling them home in the evening, by the blowing of a horn, where they were turned into the enclosures with their young, and given a little supplemental forage or grain.[100]

Because of the mildness of the winters in North Carolina, the belief arose that livestock did not require that either forage or shelter be

[96] James Westfall Thompson, *History of Livestock Raising in the United States, 1607-1860* (Agricultural History Series No. 5, United States Department of Agriculture), 1-2, cited hereinafter as *History of Livestock;* Lyman Carrier, *Beginnings of Agriculture in America* (1st. ed.), 253-259.

[97] The brand used by the Bethabara Congregation of Moravians was a "W" under an asterick, and their mark was "About one Inch Crapt off the Left Ear." Adelaide L. Fries (ed.), *Records of the Moravians,* II, 615. Laws were passed making it a felony to modify or deface a brand or mark. See also, J. D. Schoepf, *Travels in the Confederation, 1783-1784,* II, 108-109.

[98] J. D. Schoepf, *Travels in the Confederation, 1783-1784,* II, 108-109; W. K. Boyd (ed.), *Some Eighteenth Century Tracts,* 444-445. For an excellent description of a round-up and the manner in which the colonial cattle industry was conducted, see account prepared by an officer in Braddock's Army, and quoted in Archer Butler Hulbert's *Paths of Inland Commerce,* 22-24.

[99] J. D. Schoepf, *Travels in the Confederation, 1783-1784,* II, 110-111.

[100] H. J. Carman (ed.), *American Husbandry,* 240-241.

provided them in winter. Consequently, most farmers did not trouble themselves to develop meadows or to accumulate forage for winter feeding. "And hence," Schoepf noted in 1783, "most of the farmers, although they keep a number of cattle in the woods, can hardly winter one milch cow at the house."[101] Another traveler wrote in 1752:

As far as cattle & stock are concerned—it is purely their care to see to it how they get through the winter; with horses it is not better. If they survive it—they survive it! Hay they have none for there are no meadows, & corn fodder & tops do not go far. . . . This may be the reason that their horses are not much larger than English colts—and their cows the size of their yearlings.[102]

Many people, because of the neglect of their cattle, were without milk, butter, and cheese even though they possessed vast numbers of cows. This was particularly the case with farmers in the Coastal Plain where the milk of cattle, pasturing on certain swamp grasses, became so tainted as to be unfit for use. In the back country, however, there were better natural meadows and pastures, and butter and cheese became important exports from there to the Coastal Plain for either use or shipment.[103]

In view of the circumstances attending the raising of livestock in North Carolina, there can be little wonder that the losses resulting from diseases, starvation, depredations of wild animals, exposure, insect pests, theft, etc., were enormous.[104] In commenting upon such losses, one writer said that "if this were not enough for the planter raising cattle to contend with, their cows were even intercepted by the negroes and milked."[105]

There were no mules in North Carolina in 1783. The ox, because of its strength, durability, and cheapness was more commonly used as a beast of burden and draft animal than any other. When too old for work the ox was fattened for slaughter, and was just as acceptable as any beef cattle. Horses shared with the ox the draft work. The horses were of a common sort in the beginning, but, by 1783, interest was being shown in developing better animals, particularly for riding and racing purposes. They were not shod, as a general thing, and other than the well-bred saddle-horses, they had to shift for themselves just as did the

[101] J. D. Schoepf, *Travels in the Confederation, 1783-1784,* II, 121.

[102] Quoted in *N. C. C. R.,* V, 2.

[103] J. D. Schoepf, *Travels in the Confederation, 1783-1784,* II, 121; Adelaide L. Fries (ed.), *Records of the Moravians,* II, 835; L. C. Gray, *History of Agriculture,* I, 45, 205-206.

[104] Governor Arthur Dobbs reported in 1764 that North Carolina planters had lost seven-eights of their cattle as a result of a contagious disease introduced from South Carolina. A quarantine law, against the introduction of diseased cattle into the colony, was passed in the same year. *N. C. C. R.,* V, xli; L. C. Gray, *History of Agriculture,* I, 147.

[105] W. N. Franklin, "Agriculture in Colonial North Carolina," *loc. cit.,* 565.

cattle. Horse stealing was common in the state and was a felony.[106]

Large numbers of hogs were raised in the state. These, too, ran wild in the forests and swamps, feeding on the abundant mast, roots, and wild growth. In 1783, the practice was becoming more common to pen those which were to be slaughtered, and fatten them on corn. "A bushel of corn a week," Schoepf wrote, "is allowed each head, for 5-6 weeks. The amount of corn made determines the number of hogs to be fed."[107] The fattened hogs, though of a common or so-called "native" breed, brought on the hoof from three to three and one-half Spanish dollars per hundred weight in 1783. "Nowhere on the whole continent," Schoepf asserted, "is the breeding of swine so considerable or so profitable as in North Carolina."[108]

Though never so numerous or offering as much easy profit as cattle and swine, sheep were raised in the state in large enough numbers to provide considerable quantities of wool for domestic use and mutton for the planter's table. Though less capable of fending for themselves, sheep received no better attention than other kinds of livestock. John Lawson was highly pleased with the quality of both the wool and mutton produced from North Carolina sheep.[109]

Very few goats were raised in North Carolina at any time during the colonial period. John Lawson attributed the people's lack of interest in that animal to its reputation for "being mischievous to orchards and other trees. . . ."[110]

There was plenty of poultry raised in the state for domestic use. Nearly every settler owned some chickens, ducks, geese, or turkeys. These, too, were suffered generally to shift for themselves, being fed only enough to keep them near the house and from reverting to a wild state. Janet Schaw considered the methods employed in managing poultry very bad, "not a tenth part of what is hatched ever coming to perfection, tho' those that do escape their nursing come on prodigiously fast."[111] The fighting cock was about the only fowl that excited the breeder's interest. Cock fighting was a sport which attracted the patronage of high and low alike.[112]

Honey bees seem to have been native to North Carolina and to have made up part of the wild-life of every forest. Honey and beeswax be-

[106] E. W. Andrews and C. M. Andrews (eds.), *Journal of a Lady of Quality,* 280-281; W. K. Boyd (ed.), *Some Eighteenth Century Tracts,* 446; New Bern *North Carolina Magazine;* or, *Universal Intelligencer,* September 14, 1764.

[107] J. D. Schoepf, *Travels in the Confederation, 1783-1784,* II, 109-110.

[108] *Ibid.*

[109] John Lawson, *History of North Carolina,* 45. See also, W. K. Boyd (ed.), *Some Eighteenth Century Tracts,* 446.

[110] John Lawson, *History of North Carolina,* 235.

[111] E. W. Andrews and C. M. Andrews (eds.), *Journal of a Lady of Quality,* 166. See also, W. K. Boyd (ed.), *Some Eighteenth Century Tracts,* 442-443; and, L. C. Gray, *History of Agriculture,* I, 6, 208.

[112] W. C. Watson (ed.), *Memoirs of Elkanah Watson,* 261-262.

came important items of consumption and trade, and appear in nearly every computation of total exports. "Prodigious quantities of honey are found here," observed a writer in 1773, "of which they make excellent spirits, and mead as good as Malaga sack."[113]

Livestock and livestock products, particularly beef and pork, and cattle and swine on the hoof, made up a considerable proportion of the export trade of the colony. North Carolina supplies of these items were also heavily drawn upon by both contenders in the period of the Revolution. Normally, the slaughter for export sale began with the first cold season of the fall, for, as one observer put it, "then they are in their Prime of flesh. . . ."[114] Complaints arose concerning the methods of slaughtering, salting, and packing of pork and beef; and the practice of exporting cattle and swine on the hoof became more common in the late colonial period. These animals were driven principally to Norfolk, Baltimore, Philadelphia, and New York, where they were fattened, slaughtered, and sold as local products.[115] Herds of cattle and swine, moving along the roads to distant markets, became a familiar sight in North Carolina. Taverns were established to accommodate the drovers, and farmers along the way picked up extra money by selling their surplus grain and forage. The keeper of the Bethabara Diary wrote on October 20, 1774, that during September and October "more than 1000 head of cattle have been driven by here on the way to Pennsylvania."[116] Such cattle, "thin as they are," brought from three to six dollars in 1783. This was almost all clear profit, except for the expenses of the drovers.[117] Estimates vary as to the volume of this trade, however, 30,000 cattle, and from 30,000 to 50,000 hogs annually, is a conservative figure.[118]

Unlike her colonial neighbors, whose agriculture tended toward the development of larger production units and concentration on either tobacco or rice, with incidental interest in subsistence crops, North Carolina generally featured smaller producing units and placed less emphasis on the staples than on self-sufficiency. As a result, North Carolina produced surpluses of a greater variety of crops than either of her neighbors, but, in total value of exports, was far surpassed by each. North Carolina's agricultural exports included tobacco, corn, meal, beans, peas, wheat, flour, livestock, beef, butter, pork, tallow, rice, indigo, leather, hides, beeswax, flax, honey, flaxseed, and hemp.[119]

[113] W. K. Boyd (ed.), *Some Eighteenth Century Tracts*, 442-443.

[114] John Lawson, *History of North Carolina*, 45.

[115] J. D. Schoepf, *Travels in the Confederation, 1783-1784*, II, 109; *N. C. C. R.*, V, 2; L. C. Gray, *History of Agriculture*, I, 149, 210-211.

[116] Adelaide L. Fries (ed.), *Records of the Moravians*, II, 835. See also, L. Carrier, *Beginnings of Agriculture in America* (1st. ed.), 228-229.

[117] J. D. Schoepf, *Travels in the Confederation, 1783-1784*, II, 109.

[118] W. K. Boyd (ed.), *Some Eighteenth Century Tracts*, 444-445; W. N. Franklin, "Agriculture in Colonial North Carolina," *loc. cit.*, 566-567.

[119] C. C. Crittenden, *Commerce of North Carolina, 1763-1789*, 53, 59; L. Carrier,

It is impossible to determine the volume of North Carolina's agricultural exports because of the absence of accurate statistics, and the fact that a large part of the North Carolina sales were to Virginia and South Carolina merchants and counted with the totals of those colonies.[120] Based on shipments from her own ports, North Carolina's share in the total export trade of the thirteen colonies, in 1769, was only a little more than two per cent and exceeded only that of New Hampshire and New Jersey.[121] Of the totals for all the colonies for 1770, North Carolina's share was one-eleventh of the corn, one-eighth of the pork and beef, and one-twenty-fifth of the "provisions." In the same year she was a very poor third in the export of tobacco, being greatly out-distanced by Virginia and Maryland.[122] Practically no overseas exports were made during the Revolution.

The poor standing of North Carolina in relative importance as an exporter of agricultural produce may be attributed largely to her lack of suitable ports, and to the great expense incident to the transportation of goods to the markets of Virginia and South Carolina. The author of *American Husbandry*, in an effort to explain the retarded development of the colony, wrote that "the principal evil was the want of ports, of which there was not one good one in all North Carolina. . . ."[123] The need for such facilities was recognized during the colonial period. Governor Gabriel Johnston said: "I have always looked upon the want of a town with a convenient port as one of the greatest obstacles to the improvement of the trade of this country and the polishing of its inhabitants."[124] Cross Creek, later Fayetteville, was incorporated in 1762 with the hope that it would attract the trade of the back country which would then be directed to Wilmington. That trade, however, continued to seek the better markets in South Carolina and Virginia. Consequently, the problem of obtaining a good port remained to perplex North Carolinians after 1783.

Because of the slovenly and careless methods employed in processing produce for market, North Carolina products acquired an evil reputation in the colonial period which proved a handicap in trade. The colony sought to eliminate some of the basis for complaint by passing inspection laws governing the packing and crating of such products as naval stores, tobacco, beef, and pork. Every cooper was required to put

---

*Beginnings of Agriculture in America* (1st. ed.), 199; H. T. Lefler and A. R. Newsome, *North Carolina*, 112-113.

[120] Crittenden estimated that fifty per cent of North Carolina's exports, in 1775, were carried overland and marketed in Virginia and South Carolina. C. C. Crittenden, *Commerce of North Carolina, 1763-1789*, 76.

[121] *Ibid.*

[122] *Ibid.*, 73-74.

[123] H. J. Carman (ed.), *American Husbandry*, 240.

[124] Quoted in R. D. W. Connor, "Settlement of the Cape Fear," *South Atlantic Quarterly*, VI (July, 1907), 285.

his mark on each cask, barrel, or hogshead that he made. This mark, like the brands used in the livestock industry, was made of record at the county courthouse. Inspection points were designated by law and only good, merchantable produce was to be passed.[125] Because of inefficient administration, however, these laws brought little improvement in the quality of products offered for sale. This lack of appreciation of proper standards of quality on the part of agricultural producers, which developed in the colonial period, proved difficult of correction in the subsequent period.

North Carolinians, with a surplus to sell, were further handicapped by the lack of a home market. There were no towns or cities in the state in 1783 with large populations dependent upon the country-side for their food, shelter, and clothing. The population was almost wholly rural and engaged in agriculture. Therefore, it offered a very poor market for any surplus of farm products. There were times, however, when because of crop failure or other calamity one community would be in dire need of food or forage, whereas, in other parts of the state these very items were going to waste for want of a market. This resulted primarily from the poor facilities for communication and transportation.

Local prices of farm products were usually low. Whatever of competition there was was between sellers and not buyers. On the other hand, prices of imported goods were usually high. Transportation was the key to that problem, and, in 1783, the solution was not in sight.[126]

The scarcity of satisfactory money in North Carolina also retarded trade and commerce. To alleviate this situation somewhat, the prices of certain agricultural products, such as corn, hemp, rice, and wheat, were fixed by the Assembly and their use as money authorized.[127] A considerable proportion of the trade, therefore, was carried on on a barter basis. "For lack of money," according to the Salem Board Minutes for 1783, "the neighbors are bringing all sorts of products to town to exchange for things they wish, such as tobacco, wool, cotton, pieces of linen, tanned hides, and the like."[128]

Some few planters in the East owned their own boats which they used to transport their produce to markets principally in the West Indies or in the commercial cities along the coast. They usually operated as merchants on the side, bringing back cargoes of merchandise for their own use and for sale or barter to their neighbors. There was also a considerable volume of traffic, in the navigable waters along the coast, by New England traders, whom John Spencer Bassett called "marine

[125] C. C. Crittenden, *Commerce of North Carolina, 1763-1789,* 49, 51, 62.

[126] C. C. Crittenden, *Commerce of North Carolina, 1763-1789,* 86.

[127] *State Records of North Carolina,* XXV, 205-206. Cited hereinafter as *S. R. N. C.*

[128] Adelaide L. Fries (ed.), *Records of the Moravians,* IV, 1850. See also, C. C. Crittenden, *Commerce of North Carolina, 1763-1789,* 88-89; and, New Bern *North Carolina Magazine; or Universal Intelligencer,* October 26, 1764.

hucksters." On the down trip their shallow draft vessels were loaded with merchandise which they traded at various "landings" with the inhabitants for naval stores, tobacco, pork, corn, peas, flour, and other products.[129] The back country producer was not so fortunate. He was compelled either to transport his surplus products long distances overland to markets in Fayetteville, Petersburg, Norfolk, Cheraw, Columbia, or Charleston, or to barter them with the local merchant where they were worth only what that merchant was willing to give for them.

Judged by the standards of his time and place, the colonial farmer was successful. Judged by present day standards or those of his contemporaries in Europe, he was, however, a colossal waster of his resources. It might be said, too, that the abundance of resources tremendously influenced the character of the people by engendering in them habits of neglect and indolence in their way of life. Nature had been so bountiful to them and it was so easy to make a living, that the incentive to arduous labor that makes and characterizes the "good husbandman" was dulled. As a result, quite often the son in North Carolina was not nearly as good a husbandman as his father had been in England, Scotland, and Germany.[130]

As to the thoroughness of his cultivation of crops, most contemporary observers agree that the colonial farmer did a poor job. He not only did not keep his fields free from weeds and grass, but he also failed to utilize the best agricultural practices of his time. The scarcity and crudeness of farm implements and the failure to use those which were available help explain the lack of thoroughness in culture. Much of the planting done by the colonial farmer, and that in North Carolina was no exception, was in fields from which the stumps and roots of trees had not been removed. The wooden plows of that time were not strong enough to cut through such obstacles; therefore, most of the work in the fields was performed with hand implements.[131] Of these, the hoe was the most important—some farmers using it almost exclusive-

[129] John Spencer Bassett (ed.), *The Southern Plantation Overseer,* 223. See also, W. K. Boyd (ed.), *Some Eighteenth Century Tracts,* 449.

[130] E. W. Andrews and C. M. Andrews (eds.), *Journal of a Lady of Quality,* 258; R. D. W. Connor, *Rebuilding an Ancient Commonwealth,* I, 38-39; A. O. Craven, *Soil Exhaustion,* 19-23. One contemporary critic said the abundance of land in North Carolina "spoiled" the farmer by enabling him to gain the necessities of life "with so much ease that an accurate and industrious cultivation appears useless." H. J. Carman (ed.), *American Husbandry,* 248, 315.

[131] A traveler through the colony in 1753 spoke of going 140 miles without seeing a single plow or wagon. Quoted in L. C. Gray, *History of Agriculture,* I, 194-195. For descriptions of plows used by the colonial farmers, see L. Carrier, *Beginnings of Agriculture in America* (1st. ed.), 263-265. John Lawson said: "I must confess that I never saw one acre of Land Managed as it ought to be in Carolina, since I knew it. . . ." John Lawson, *History of North Carolina,* 42. Another critic of farming said that "There is no greater defect in the husbandry of this province than the foulness of the crops with weeds, & c. . . ." H. J. Carman (ed.), *American Husbandry,* 259.

ly in tilling their land.[132] The axe shared the position of greatest importance with the hoe. Scythes, reaping hooks, flails, forks, mattocks, carts, harrows, and wagons were also important. Each farmer was somewhat a jack-of-all-trades and usually made his own implements. Needless to say, the implements were usually very crude. Most were made of wood, since what little iron was to be had was of poor quality and quite brittle.[133]

Although the colonial farmer adopted native plants and fruits and introduced new ones, he made practically no attempt to improve their quality or that of his livestock by either culture or selection. Exception may be had in this respect for tobacco, rice, and indigo, all of which were either improved in quality or by methods of processing during the period. These were primarily commercial crops and their improvement was impelled, to a considerable extent, by standards of quality fixed by law or maintained in the markets in which they were sold.[134]

In the absence of any law requiring the fencing in of livestock, the farmer was compelled to fence his fields. The fences were usually constructed at the time the field was cleared for planting, and, though there was abundant growth which might have been used for live fences, they were more commonly "made of a set of logs laid zagly close over each other."[135] Most contemporary critics of the colonial farmer remarked upon the state of disrepair and inadequacy of the fences. The author of *American Husbandry* wrote that "their fences are extremely incomplete, and kept in very bad order . . . this is another evil occasioned by plenty of land; they will grasp at more than they have money to cultivate . . . and then are forced to manage it in a slovenly manner."[136] This critic thought the North Carolina farmer was more neglectful of his fences than the farmers in any of the English colonies, not even fencing his rice and tobacco fields "with the care and attention which in England is bestowed on the least valuable fields."[137]

Another defect of North Carolina agriculture, which was particularly noticeable to foreign observers, was the almost universal neglect of, and failure to use, available manures. This threw an unnecessary burden on the soil and hastened the day when it would be exhausted. Undoubtedly, as one critic said, no error in agriculture could cause worse consequences than failure to be "sufficiently solicitous about manure. . . ."[138] The waste resulting from this neglect was particularly great in North

[132] E. W. Andrews and C. M. Andrews (eds.), *Journal of a Lady of Quality*, 163; L. C. Gray, *History of Agriculture*, I, 194-195.

[133] E. L. Bogart, *Economic History of the American People* (2d. ed.), 54.

[134] *Ibid.*, 49.

[135] E. W. Andrews and C. M. Andrews (eds.), *Journal of a Lady of Quality*, 163.

[136] H. J. Carman (ed.), *American Husbandry*, 190.

[137] *Ibid.*, 248-259.

[138] *Ibid.*, 105.

Carolina in view of the vast herds of cattle maintained in the state. Actually a prejudice existed against using manures under crops. Janet Schaw's sister-in-law was shocked at the idea of manuring the fields, "and declared she never would eat corn that grew thro' dirt."[139] J. D. Schoepf observed in 1783 that "until the earth is quite exhausted there is no taking in a new piece of land, for it never suits their fancy to better the old by dunging."[140] The neglect of other means of retaining soil fertility was equally as great.[141]

Several proposals for reform of agriculture in colonial North Carolina were made but nothing came of them.[142] The author of *American Husbandry* even suggested the establishment in each colony of a large experimental plantation, which would be maintained at public expense. He not only outlined the manner in which experimentation should be conducted, but also prepared several estimates as to the probable expenses and income of such an institution over a period of years. His remarks in connection with this proposal constitute the most constructive appraisal and criticism of agriculture in colonial North Carolina.[143]

These proposals of improvements were not carried out, however. The Revolutionary War came and with it the end of the colonial status. What effects independence would have on the course of agriculture in North Carolina, no one could foresee. Elkanah Watson probably voiced the prevailing sentiment concerning North Carolina when he wrote in 1786: "Perhaps no state had at that period performed so little to promote the cause of education, science, and arts, as North Carolina. . . . The lower classes of that region were then in a condition of great mental degradation."[144] The picture was not entirely dark, however. The same observer found areas around Hillsborough, Halifax, and along the lower Cape Fear, which were "in an advanced state of agricultural improvement, and embraced a very genteel society."[145] That general change and improvement would be slow and difficult, there could be no doubt. Already in some communities, one agricultural operator might be found employing the most advanced techniques of production; whereas, his next door neighbors remained entirely uninfluenced and held doggedly to the outmoded practices of their fathers.

[139] E. W. Andrews and C. M. Andrews (eds.), *Journal of a Lady of Quality*, 160.

[140] J. D. Schoepf, *Travels in the Confederation, 1783-1784*, II, 138-139.

[141] E. W. Andrews and C. M. Andrews (eds.), *Journal of a Lady of Quality*, 163.

[142] John Lawson, *History of North Carolina*, 42; H. J. Carman (ed.), *American Husbandry*, 193-194, 196-197, 226-234, 250-256.

[143] H. J. Carman (ed.), *American Husbandry*, 193-194, 196-197, 226-234, 250-256.

[144] W. C. Watson (ed.), *Memoirs of Elkanah Watson*, 253.

[145] *Ibid.*

CHAPTER II

# POST-REVOLUTIONARY DEVELOPMENTS, 1783-1820

The period from 1783 to 1820 witnessed the beginnings, perhaps, of every development affecting the progress of agriculture in ante-bellum North Carolina. There appeared in this period in nearly every county of the state one or more leaders who might have been classified as "improving" farmers. Based upon observations gathered by correspondence, travel, reading, and experimentation, these men appraised the agriculture of North Carolina in the light of the best information of the times, and in many particulars found it wanting. Without fanfare or ostentation, working as individuals without the backing of any group or organization, and quite often being the objects of censure and ridicule by their neighbors, these men set in motion new techniques of operation and schools of thought in the realm of agriculture which were designed to counteract the evil effects of the old system. That progress was extremely slow, and that the great mass of farmers was scarcely conscious that changes were being made, goes without saying. North Carolina, in that day, had a reputation for backwardness which the facts do not belie. Significant developments were made, however, and these laid the groundwork for the more successful agriculture of the late ante-bellum period. The purpose of this chapter is to survey the period from 1783 to 1820 and to note those developments. In many respects they are found to reflect the differences between the pioneer and succeeding stages in our economic and social development.[1]

The war for independence brought widespread destruction of property, disorder, and disruption of normal routine to the population of North Carolina. Scarcely a community in the state escaped, for, in addition to the more formalized campaigning of the regular forces, the Revolution was a civil war in which, because of divided sympathies, neighbor was pitted against neighbor. The results of the civil strife were particularly bad in the section where the two parties were about equal in size. That was the case in Stokes and Surry counties, for example, where, according to a contemporary author, some leaders made themselves notorious by their plundering and marauding conduct. "No principal of moral duty," he wrote, "seemed to have existence among these people at that time—The strongest law prevailed and made the rule of right."[2]

[1] A. O. Craven, *Soil Exhaustion*, 107-109.

[2] Albert Ray Newsome (ed.), "Twelve North Carolina Counties in 1810-1811," *North Carolina Historical Review*, VI (July, 1929), 302.

Grain, meat, and livestock of all sorts were in great demand during the war and many levies, either legal or illegal, were made by the contending parties upon the farmers of the state for the supply of those items. In the latter stages of the war, North Carolina was a principal source of livestock and livestock products for both sides.[3] As salt for the curing of meat became scarce, a larger proportion of the pork and beef product of the state was driven to markets in Charleston, Baltimore, and Philadelphia, where it was corn-fattened before slaughter. Otherwise, practices employed in livestock raising in the state changed very little during the war and for several decades thereafter. The tendency of the industry to move westward as the area of settlement increased had set in before the Revolution.[4]

The war brought a disruption of commerce which had serious effects on the agriculture of the state. The British government, of course, ceased bounty payments, and prohibited shipments of tobacco and naval stores to Great Britain and provisions, livestock, fish and lumber products to the British West Indies. This necessitated finding new markets for these products, and resulted in shortages of those items which had previously been imported.[5]

Of far more immediate consequence to the mass of people in the state than the loss of bounties and markets was the shortage of salt which resulted from the war. This shortage was most severely felt in the back country where not enough salt was available for the processing of beef and pork products and for household usage. The price of this article skyrocketed in most of the states. Toward the end of the war, however, new sources of supply were found and the supply more nearly met the demand.[6]

Pending the establishment of new trade connections, the commercial planters of rice, indigo, and tobacco were in a depressed condition as a result of the war. Tobacco was less badly affected than the others because the state not only accepted it in payment of taxes but purchased large quantities for export in exchange for needed manufactured goods.[7] An increase in trade with the non-British West Indies, as well as the opening of markets in Northern Europe for tobacco and in Southern Europe for rice offset, somewhat, the acuteness of the depression. Rice

[3] Allan Nevins, *The American States During and After the Revolution, 1775-1789*, 382; L. C. Gray, *History of Agriculture*, II, 583-584.

[4] J. W. Thompson, *History of Livestock*, 77.

[5] C. C. Crittenden, *Commerce of North Carolina, 1763-1789*, 126-127, 131-133. One authority has estimated that between 1780 and 1787, after the import of provisions from the continental colonies was prohibited, over 15,000 slaves died in the British West Indies of famine or disease resulting from the scanty and unwholesome diet to which they were reduced. E. L. Bogart, *Economic History of the American People* (2d. ed.), 216-217.

[6] L. C. Gray, *History of Agriculture*, II, 584; C. C. Crittenden, *Commerce of North Carolina, 1763-1789*, 137.

[7] C. C. Crittenden, *Commerce of North Carolina, 1763-1789*, 132-133; U. B. Phillips (ed.), *Plantation and Frontier Documents: 1649-1863*, I, 85.

culture, too, was aided by the invention in 1778 of a water mill for use in cleaning the grain.[8]

Shortages of sugar, molasses, and rum, which had previously been imported from the British West Indies, were acutely felt. Inability to get the required supply of rum gave encouragement, however, to the establishment of distilleries in the state for the manufacture of whiskies and brandies. This industry became widespread and continued to expand after the war as farmers found an universal market for their surpluses of grain and fruit when in liquid form. According to an estimate made in 1811, there were 5,381 distilleries in the state which produced an annual output valued at $700,105. This volume of production gave distilling rank as one of the leading industries in the state.[9]

After the war England kept the ports of her West Indian colonies closed to American ships, but British ships moved busily between those ports and the United States seeking to renew the old trade connections. Provisions again were in great demand, and, by 1788, the value of shipments of corn, peas, flour, and fish exceeded that for any pre-war year of record. Tobacco, too, no longer restricted to the British market, by 1788, showed an appreciable increase over recorded pre-war shipments. Though the export business had been slack in 1783, by 1786, it was fairly prosperous.[10] In comparative importance, however, the trade of North Carolina continued relatively small. The total value of exports from the state for the twelve months' period ending September 30, 1802, amounted to $659,390 and only exceeded those of New Hampshire, Vermont, New Jersey, Delaware, and Kentucky for the same period. Virginia and South Carolina were far in advance with exports valued at $4,660,361 and $10,639,365 respectively.[11]

Those engaged in self-sufficing agriculture, and in North Carolina this group was in great majority in the population, were but little affected by the war and the resultant disruption of trade and commerce.[12] The difficulties of transport and the absence of markets were such that they sold little and bought only essentials, such as salt, ammunition, and hardware.[13] A contemporary account of the trade of the back country in 1802 indicates that it represented the same items, except for the increased shipments of cotton and tobacco, and followed the same channels as in the colonial period.[14]

[8] E. L. Bogart, *Economic History of the American People* (2d. ed.), 210; L. C. Gray, *History of Agriculture,* II, 579-580.

[9] Raleigh *Star,* February 21, 1811; E. L. Bogart, *Economic History of the American People* (2d. ed.), 210-211.

[10] C. C. Crittenden, *Commerce of North Carolina, 1763-1789,* 157-161. For a brief account of adjustments made by Great Britain relative to her trade with the United States in 1783, see L. C. Gray, *History of Agriculture,* II, 600.

[11] Raleigh *Minerva,* February 8, 1803.

[12] E. L. Bogart, *Economic History of the American People* (2d. ed.), 209.

[13] U. B. Phillips, *American Negro Slavery,* 155.

[14] Reuben Gold Thwaites (ed.), *Travels West of the Alleganies Made . . . in*

Because of the deep-seated hatred of taxation in all forms, the North Carolina state government had sought to finance the Revolution by several different issues of paper money. As more and more of this money was issued specie went into hiding, with the result that violent fluctuations occurred in the level of prices. By 1781, in purchasing power, seven hundred and fifty paper dollars were equal to only one dollar in specie.[15]

In the absence of a good currency system, a considerable portion of business transactions was carried on on a barter basis. Merchants advertised that they would take in exchange for their wares such items as "likely negroes," wheat, corn, peas, tobacco, deerskins, hemp, flax, beeswax, tallow, and naval stores.[16]

The state made little attempt to change this condition of affairs, and for several decades after 1783, the financial situation in North Carolina was unstable. After the chartering of the Bank of Cape Fear and the Bank of Newbern in 1804 the money situation improved somewhat. In 1810, the State Bank of North Carolina was chartered—with its central bank to be located in Raleigh and branches at Edenton, Newbern, Wilmington, Fayetteville, Tarboro, and Salisbury.[17]

In addition to cash transactions and the widespread use of barter, much of the business of the day was done on credit. A contemporary described conditions in Rockingham County in 1810 as follows:

> here, as in most parts of the back Country, Commerce and Trade is engrossed by a few enterprising individuals, who generally keep the People in debt to them; too many of our Farmers and planters, owe the proceeds of their crop to the Merchant before it is put into the ground, which is the principal cause of the Scarcity of cash so often felt, and complained of in this part of the State.[18]

Though increased production was stimulated during the Revolution by the British blockade, the culture of short staple cotton continued to be for home use only until after the invention of the cotton gin in 1793.

---

*1802 by F. A. Michaux. . . ,* 300. See also C. C. Crittenden, *Commerce of North Carolina, 1763-1789,* 164-165.

[15] William Kenneth Boyd, "Currency and Banking in North Carolina, 1790-1836," *Historical Papers* published by the Trinity College Historical Society, Series X, 1914, 56-65; C. C. Crittenden, *Commerce of North Carolina, 1763-1789,* 134-135; L. C. Gray, *History of Agriculture,* II, 586-589.

[16] Edenton *State Gazette of North Carolina,* December 24, 1789; C. C. Crittenden, *Commerce of North Carolina, 1763-1789,* 164.

[17] W. K. Boyd, "Currency and Banking in North Carolina, 1790-1836," *loc. cit.,* 68-69. For an able account of banking developments in North Carolina during the period 1804 to 1836, see *ibid.,* 70-86. For a more comprehensive treatment of banking in ante-bellum North Carolina, see Brantson Beeson Holder, "The Three Banks of the State of North Carolina, 1810-1872." Unpublished doctoral dissertation at the University of North Carolina, 1937.

[18] A. R. Newsome (ed.), "Twelve North Carolina Counties in 1810-1811," *N. C. H. R.,* VI (July, 1929), 300.

This machine revolutionized the culture of that crop in North Carolina as it did elsewhere.[19] With it, one man could clean three hundred and fifty pounds of cotton in a day; whereas, by using the hand method, he formerly cleaned only one or two pounds.[20] This fact, together with the realization that no expensive equipment or particularly skilled labor was required in its culture, led many non-slaveowners as well as commercial planters to undertake the growing of cotton. This was especially the case in the Piedmont for it was also known that cotton could withstand the difficulties of transport to market without losing much of its value.

The number of farmers engaged in the culture of cotton increased rapidly, and as prices continued high, many devoted their profits to the purchase of slaves for expansion in operation.[21] As early as 1802, a traveler noted a German farmer in Lincoln County who had 150 acres in cultivation and was specializing in cotton production. This man owned his own gin which was driven by water power and operated by his slaves. Unlike most farmers in that section, this farmer "dunged" his fields annually.[22]

John Steele, of Rowan County, was an early large scale cotton planter in the Piedmont. In 1805, he was planting his 800-acre plantation, "Lethe," located on the Yadkin River below Trading Ford, and shipping his product by wagon to market in Petersburg. He also owned a 900-acre plantation named "Lombardy," which was located just out of Salisbury. In 1814, he owned 115 slaves, valued at $32,250. Steele and his family resided at "Lombardy" and entrusted the management of "Lethe" to an overseer.[23]

Governor Benjamin Williams, writing to General Henry William Harrington of Richmond County from Raleigh in 1800, thanked him for a wagon load of cotton seed and stated his intention of planting a crop since it "will be the safest crop for a beginner." Governor Williams' plantation was in Moore County. He instructed his overseer to "pitch" his crop entirely in hills prepared in squares four feet apart.[24] From

[19] U. B. Phillips (ed.), *Plantation and Frontier Documents,* I, 85. For an account of the invention of the cotton gin, see Jeannette Mirsky and Allan Nevins, *The World of Eli Whitney,* 61-80. Whitney's short description of the gin is found in Appendix "A" of that volume.

[20] Emory Quinter Hawk, *Economic History of the South,* 234.

[21] U. B. Phillips (ed.), *Plantation and Frontier Documents,* I, 93. In a committee report made to the North Carolina House of Commons in 1828, Charles Fisher referred to cotton as "our other great staple" and estimated that the state had been producing an annual crop of about 80,000 bales. Report reproduced in *Report on the Geology of North-Carolina. . . . ,* Part III, 46.

[22] R. G. Thwaites (ed.), *Travels West of the Alleganies Made . . . in 1802 by F. A. Michaux. . . . . ,* 292.

[23] Henry McGilbert Wagstaff (ed.), *The Papers of John Steele,* I ,425 n; II, 725.

[24] Henry McGilbert Wagstaff (ed.), "The Harrington Letters," *James Sprunt Historical Publications,* XIII, No. 2 (1914), 21. General Harrington resided at

this crop, which occupied forty-two acres, 3,200 pounds of clean cotton was harvested. Russel's gin was being used on this farm and giving satisfactory service.[25]

From 1793 onward, the "great cry about cotton," to use an expression frequently found in the correspondence of that time, resulted in an enormous increase in the acreage devoted to its culture. By 1811, there were 739 gins in operation in the state which cleaned annually 2,475,810 pounds.[26] Not only were profits good from sales abroad but, because of the South's meager success in the culture of wool, cotton came more and more into demand by her domestic industries in the early nineteenth century.[27] John Taylor, of Caroline, reflected the high esteem in which cotton was held by the leading agriculturists of that time, when he wrote in his *Arator* in 1809 that "There is no article of our agriculture, the raising of food excepted, more worthy of attention."[28]

Though livestock raising became relatively less important in North Carolina in this period, several significant developments occurred. Mules made their first appearance in the state, and, according to an observer in 1810, they were "growing into use" and were "much approved where fairly tried."[29] In 1796, the "noted" jack, "Tippo Saib," described as thirteen hands high and large bodied, was advertised as standing for the season at Trenton.[30] The mule proved of great value on the farm and was increased in numbers rapidly, but never completely displaced the horse and the ox.

The extravagant praise of Merino sheep, which were first imported into the United States in 1801, early attracted the interest of North Carolina sheep raisers. W. B. Alexander, of Mecklenburg County, "after much importunity," obtained a pair of full-blooded Merinos from Chancellor Robert Livingston, one of the original importers, which

his plantation, "Beausejour," on the east side of the Pedee River in what is now Richmond County. He had acquired this property in 1776, which, according to the editor of these papers, was "a typical southern slave plantation until the Civil War." He also had a plantation near Cheraw, South Carolina. Described by one of his correspondents as "the first farmer in the State," General Harrington early shifted to the large scale culture of cotton and his advice on this and other agricultural subjects was sought by other planters. Though not to be classified as a "scientific" farmer, General Harrington attempted to build up his lands, by manuring and experimentation with various grasses. *Ibid.*, 6, 8-9, 18-20.

[25] *Ibid.*, 23-24.

[26] Raleigh *Star*, February 21, 1811.

[27] Charles Fisher, in a report to the legislature in 1828 urging the establishment of cotton and woolen factories in North Carolina, estimated that one-fourth of the state's annual cotton crop, or approximately 20,000 bales, was being consumed in domestic manufacturing in the state. Report reproduced in *Report on the Geology of North-Carolina. . .*, Part III, 45-47.

[28] Reprinted in *Farmer's Register*, VIII (December, 1840), 761.

[29] A. R. Newsome (ed.), "Twelve North Carolina Counties in 1810-1811," *N. C. H. R.*, VI (April, 1929), 184.

[30] Newbern *North-Carolina Gazette*, April 6, 1796.

were delivered to Charlotte in 1809.[31] In 1810, George Pollock brought a Merino ram to Newbern for which he had paid $350 in Philadelphia.[32]

Like those of other sheep raisers, Alexander's flock suffered from the depredations of dogs and foxes. "If our Legislature," he wrote, "are desirous to encourage the raising of this very valuable animal, they will certainly lay a tax—upon dogs. The expense and losses occasioned by the superfluous dogs in this state would victual an army."[33] Nevertheless, Alexander announced the next year that by August he would have sixty half-blood Merino lambs for sale.[34] "Poor Richard" observed in 1813 that wool was bringing fifty cents a pound and many farmers were becoming "more attentive to encreasing their Flocks of Sheep but were being checked in that design by sheep-killing dogs."[35] "J. B." was the author of an article in 1813 strongly advocating action by the legislature to curb the dog menace to this industry.[36] The appeal for state aid in this respect continued until the legislature passed such a measure in the late 1850's.

As for horned cattle and swine, most accounts agree that the herds to be found in North Carolina were smaller than those of the colonial period. Every expansion of settlement and opening of new fields reduced the "common" foraging grounds and the opportunity of landless people to raise livestock.[37] To a considerable extent, livestock were suffered to shift for themselves, receiving very little attention as to their feeding, shelter, health, or breeding.[38]

In 1795, the legislature passed an act forbidding South Carolina and Georgia cattle drovers to drive their herds into North Carolina at any time in the year between April 1 and November 1 under penalty of forty shillings per head. This act also forbade North Carolinians, during the same months, to drive cattle from a place where the soil was sandy and shaded by long leaf pines into or through the high lands of the state where the soils and timber were different. All persons driving cattle through the state thereafter were required to have a certificate indicating what kind of region the cattle came from and that they were free from any infectious distempers.[39]

A few of the more intelligent farmers began to give attention to the

[31] Raleigh *Star,* June 28, 1810.
[32] *Ibid.,* May 31, 1810.
[33] Raleigh *Star,* June 28, 1810. This issue carried the following recipe for preserving lambs from foxes: "Mix equal quantities of tar, train oil, and block brimstone, dress the lambs as soon as possible with as much as will lie on a half crown of this ointment."
[34] *Ibid.,* July 12, 1811.
[35] *Ibid.,* April 30, 1813.
[36] *Ibid.,* November 12, 1813.
[37] Francis Grove Morris and Phyllis Mary Morris, "Economic Conditions in North Carolina About 1780," *N. C. H. R.,* XVI (July, 1939), 316-317.
[38] Buford Rowland (ed.), "Some Observations on the Low Parts of North Carolina," *N. C. H. R.,* XV (April, 1938), 157.
[39] *Laws of the State of N. C.* (1795), I, 783-784.

care and quality of their livestock. With regard to improvement of quality, more interest was taken in the horse than in any other animal. Practically all the state newspapers carried announcements of the "stands" of stud horses along with descriptions of their pedigrees and records. The principal interest was in a better race horse, saddle horse, or carriage horse, rather than a better work animal for the farm.

George W. Jeffreys, North Carolina's leading agricultural reformer, scolded those farmers who were interested in the improvement of horses only and asserted that the "great number of horses in our country is a serious and sore evil; and the wasteful manner in which they are fed, makes it much more grievous." He advocated dispensing with every superfluous horse, of which, he said, "the aggregate number . . . is great."[40] As for other animals, he wrote: "The possession and proper management of live stock, are perhaps the most important objects which claim the attention of the farmer, as upon these the fertility and prosperity of farms may be said to depend."[41]

Jeffreys was the first North Carolinian to recognize the advisability, in general farming, of effecting a balance on the farm between culture of crops and livestock. ". . . there is," he wrote in 1819, "such a connection between the cultivation of ground, and breeding, rearing, and fattening cattle, sheep and other domestic animals, that a man will make but an indifferent figure in rural affairs, if he does not understand the latter as well as the former."[42]

Though very little was at first accomplished toward the improvement of livestock, other than horses, by selective breeding, the ground work was laid for considerable progress in the late ante-bellum period. It was quite significant, also, that the more intelligent farmers were becoming interested in the culture of grasses, a great part of which crops were used for either grazing or supplemental feeding of livestock.

"A Richmond County Farmer" wrote in 1810 that he was "soiling" his livestock with a mixture of wheat straw and freshly cut red or Dutch clover. ". . . it all makes the best hay that cattle ever eat," he said. He was also growing red top and "East India light mountain grass," which he considered easy of culture and good feed.[43]

This correspondent was experimenting with the use of plaster of Paris and gypsum on his grass crops. He washed his seed grain, and, while still wet, shook it together with plaster until the seed was encrusted.

[40] (George W. Jeffreys), *A series of Essays on Agriculture & Rural Affairs; in Forty-Seven Numbers,* 195. Jeffreys wrote these essays under the pen name, "Agricola." As to the importance of livestock breeding, he wrote: "Improving the breed of live stock, is as necessary for the farmer, as the proper cultivation of a field for wheat, corn, or any other crop." *Ibid.,* 202. Cited hereinafter as *Essays on Agriculture.*

[41] *Ibid.,* 191.

[42] (George W. Jeffreys), *Essays on Agriculture,* 202.

[43] Raleigh *Star,* August 2, 1810.

The seed was then planted. On some fields, he planted his seed without treating it and then broadcast over the land the same quantity of gypsum as he had planted of seed. In all, he tried one or the other of these practices on eleven different kinds of grasses, but apparently was not very well pleased with the results.[44]

Jeffreys reflected the increasing interest of the better farmers in the culture of grasses by strongly advocating their use on all cleared lands which were not under the plow. He would make such a practice a maxim in agriculture. He emphasized the fact that grass would hold land together and prevent washing. Grass would furnish shade in summer and warmth in winter, and, he wrote, "is at the same time drawing supplies of manure from the atmosphere through the vegetable process, to be given to the earth through the agency of the plough."[45]

Some few farmers began to give more attention to the culture of grasses. "Poor Richard" wrote that rye should be more popular as a grain crop in North Carolina. He praised it as the most nutritious feed for livestock; it also made good bread, good whiskey, and was good for the soil on which it was grown. Like the author of *American Husbandry* of an earlier date, this correspondent urged the farmer to plant all worn out lands in grasses, and recommended rye for that purpose, rather than "destoying his wood lands at vast expense and labour, so as to procure new ground."[46] Though the interest did not become widespread, there was a beginning of experimentation in the culture of the grasses in North Carolina which continued throughout the ante-bellum period.

As for new crops, "A German" wrote in 1810 that the Chinese radish had been successfully cultivated in North Carolina since 1807. This crop was grown principally for its seed from which an oil somewhat similar to olive oil was extracted. Its culture, however, does not appear to have become widespread or important.[47]

In the absence of an adequate home market, most contemporary writers agreed with John Taylor that "The best markets of our agriculture lie in foreign countries. . . ."[48] Conditions in Europe between 1789

[44] *Ibid.* In this article, the author also set forth his method of curing red or Dutch clover.

[45] (George W. Jeffreys), *Essays on Agriculture*, 26-27. In these essays, Jeffreys recommended the planting of red clover, timothy, orchard grass, tall meadow oat, Guinea grass, herds grass, and red top. *Ibid.*, 87-103. He devoted an entire essay to red clover. *Ibid.*, 118-122.

[46] Raleigh *Star*, October 16, 1812. "Poor Richard," in his desire to "render service to the public and guide the attention of Farmers to their best interests," enclosed to the editor of the *Star* an article, taken from the *Weekly Register*, which strongly advocated an increase in sheep raising.

[47] Raleigh *Star*, May 3, 1810. The editor offered free seed of this plant to those who would apply to his office. This is the first instance, noted by the author, of a newspaper editor rendering this sort of service to the farming interests in the state. The practice of distributing free seed became an important contribution of ante-bellum editors.

[48] From reprint of the *Arator* in *Farmer's Register*, VIII (December, 1840), 709.

and 1815 were such that unprecedented demands were made upon American agriculture to supply farm products which Europeans were prohibited from raising either by nature or the wars which beset them. France was not nearly self-sufficing in the cereals in 1789. In November of that year, the French government announced that bounties would be paid French merchants for the importation of wheat, rye, barley, and the flour thereof "from the different parts of Europe, or of the United States of America. . . ."[49]

England and other states in Europe imported grains, tobacco, and cotton from the United States. Between 1793 and 1807 the price of flour in the Philadelphia market averaged $9.12 per barrel; whereas, for the nine years previous to 1793 it averaged only $5.41, and for the nine years after 1807, only $5.46.[50] After peace came to Europe in 1815, England enacted a Corn Law which practically denied the admission of wheat and flour into that country, but the demand for cotton continued.[51] North Carolina agriculture was affected by all these conditions and events.

The embargo imposed by the United States in 1807 met with vigorous opposition by some Americans. "Philo Agricola" complained in the *United States Gazette* that it would cost the farmers millions of dollars. *"Hardly a poor man,"* he wrote, *"but will feel he has lost the value of a cow, of a horse, perhaps of a small piece of ground, where he might sit down in ease; all in consequence of a ruinous embargo. . . ."*[52]

The embargo along with non-intercourse and the War of 1812 brought depression to the commercial planters who were dependent upon the European market. This condition stimulated self-sufficing farming and the production of hemp and flax for use in domestic manufacture, which was flourishing in the state at that time.

Wightstill Avery, of Burke County, wrote of a farmer in Haywood County who planted hemp in with his corn and cultivated both at the same time. From twelve acres, planted in this manner, he harvested eighty bushels of good hemp seed for which he received $3 per bushel or $20 per acre in addition to the hemp and corn.[53]

Prices for agricultural produce went down during the War of 1812 and those for manufactured goods rose. Ann Nessfield, of Fayetteville, observed in 1813 that "our country seems to be in a wretched state,

[49] Edenton *State Gazette of North-Carolina,* April 3, 1790.

[50] E. L. Bogart, *Economic History of the American People* (2d. ed.), 235.

[51] *Ibid.,* 241.

[52] Reprinted in Raleigh *Minerva,* September 29, 1808.

[53] Raleigh *Star,* March 8, 1810. This issue, like most of the state papers of that time, carried an article on the culture of hemp. Like most articles on agricultural subjects appearing in North Carolina newspapers in the ante-bellum period, this one was taken from another paper. This particular article was taken from the *Massachusetts Spy.* With very few exceptions, when it came to writing on such subjects, North Carolina planters as well as farmers were perhaps less articulate than those of other states.

every article of consumption doubled in price."[54] On the other hand, the widow of General Henry William Harrington referred to above, wrote in 1812 as follows:

> In answer to your question about the settling of the estate, I can only say that I hope to get through at last; but the present dull price of "produce" will I fear make the negroes hire for very little and the lands will rent in proportion, but . . . I have no fear of being obliged to sell any of the negroes . . . unless times grow more desperate than at present.[55]

The appearance in several North Carolina newspapers, notably the Raleigh *Star,* of sections devoted to "Rural Oeconomy," "Agriculture," "Rural Affairs," etc., indicates that more intelligent concern for the agricultural interests was beginning to be felt. In these sections, various aspects of agriculture and rural living were discussed either in articles by local contributors or in reprints of articles taken from other papers. Unfortunately, the latter source was more prolific. Quite often, the selections of such articles were unwisely made, in that subjects were presented to North Carolina readers which were inapplicable because of climate and soil differences. The editors, too, in their zeal for the cause of agriculture were often too hasty in giving their stamp of approval and publicity to new implements and practices just because they were new. More will be said on this score later.[56]

The articles carried had to do with such subjects as the following: how to make Parmesan cheese, how to preserve bacon from bugs and worms, culture of senna, how to get rid of cheat in wheat, recipes for spirits of all kinds, causes of dry-rot in wood, how to cure a wound on a tree, how to make sweet oil, preserving wheat from weevils, new methods of raising strawberries, profitableness of hemp culture, planning and location of farm buildings, care of grounds, care of orchards, and the experimentation of Sir Humphrey Davy in breaking down potash, soda, and lime into their constituent elements.

On February 5, 1810, the Cape Fear Agricultural Society was founded in Wilmington to promote and improve farming in its various branches. This society planned to give awards for "all improvements in

[54] H. M. Wagstaff (ed.), *Papers of John Steele,* II, 714.

[55] H. M. Wagstaff (ed.), "The Harrington Letters," *loc. cit.,* 38.

[56] One of the most revealing and, perhaps, accurate accounts of non-political conditions in North Carolina in this period is to be found in the Thomas Henderson "Letter Book." Henderson, editor of the Raleigh *Star,* solicited the articles which comprise this source from well informed individuals of various counties. Though lacking in balance, completion, and equality in merit for the twelve counties upon which articles were written, they offer a variety of contemporary views on many aspects of the agricultural situation in the state in 1810-1811. Too, they contain much of events and trends concerning which the authors had firsthand information. See A. R. Newsome (ed.), "Twelve North Carolina Counties in 1810-1811," *N. C. H. R.,* V (October, 1928), VI (January, April, July, 1929).

agriculture or in any species of domestic economy."[57] Soon, the editor of the *Star* was urging the formation of such societies in every county of every state.[58] Though no such comprehensive response was made to this proposal, it is significant to note that the movement toward the organization of agricultural societies in North Carolina ran parallel with, and not behind except in degree of permanence and success, the "Berkshire Plan" of such societies then beginning to spread throughout the states north of Pennsylvania. The annual fair, however, which was a main feature of the northern societies, did not make its appearance in North Carolina until later. This was unfortunate, for fairs, when properly promoted, were far less formalized than the societies and proved to have much greater appeal to "dirt" farmers than expositions on theoretical science in agriculture.[59]

Another significant development of the period was the beginning of publication of almanacs in the state. Agriculturists who, wedded to the old practices, heaped severe ridicule upon "book farmers" found it within their conscience to subscribe for an almanac. Furthermore, the almanacs also carried articles on agricultural subjects.

> Except for the Bible, it was probably the most universally read publication in the American farm home during the eighteenth century. Although intended primarily to record astronomical data, it readily became a depository for all manner of miscellaneous rural information. Frequently verse, jokes, anecdotes, riddles, essays, 'cures,' recipes, and interesting bits of general information appeared.[60]

To John Taylor, of Virginia, and to his disciple, George W. Jeffreys, of North Carolina, must go a considerable portion of credit for initiating reform in agriculture in North Carolina. Taylor's *Arator* was first published as separate essays in a newspaper in 1809 or 1810. In 1812, the essays were collected and published in book form, which, by 1818, was in its sixth edition.[61]

Edmund Ruffin, a later and greater agricultural reformer of Virginia, criticized Taylor for his ignorance of the chemistry of soils, for having spoken "contemptuously" of marl and lime as manures, and for improperly recommending for general use practices which were suitable only on soils similar to his own.[62] Yet Ruffin recognized the *Arator* as containing so many fundamental truths about farming that he reprinted the sixth edition in his *Farmer's Register* in 1840.

According to Ruffin, the *Arator* became "an immediate sensation" because of a "fortunate combination of circumstances" which existed at

[57] Raleigh *Star,* March 22, 1810.
[58] *Ibid.,* January 31, 1811.
[59] Albert Lowther Demaree, *The American Agricultural Press, 1819-1860,* 9. See also, Wayne Caldwell Neely, *The Agricultural Fair, passim.*
[60] A. L. Demaree, *American Agricultural Press,* 9.
[61] *Farmer's Register,* VIII (December, 1840), 704.
[62] *Farmer's Register,* VIII (December, 1840), 703-704.

the time of its issue. It appeared at a time when men of wealth and intelligence, though vitally interested in its success, were neglecting to improve their agricultural practice even though the evidences of evil in those practices were to be seen on every hand. Benefited by the European wars, even "the most wretched agricultural practice" was rewarded with high profits up to 1807. After that date, depredations on American commerce and the "singular defensive policy of this government" reduced the profits to American farmers to less than fifty per cent of former totals. Such, wrote Ruffin, were the conditions when the *Arator* was first published, and because Taylor's views were so highly esteemed, "even his most erroneous views and opinions were received as readily and universally as the most sound and unquestionable."[63] Some of Taylor's disciples, Ruffin said, even went beyond the positions of their teacher and many errors of practice and consequent losses were made.[64] Be this as it may, there is no doubt that Taylor gave a tremendous impulse to agricultural reform which was felt in North Carolina as well as in Virginia.

George W. Jeffreys, North Carolina's leading agricultural critic and reformer, reflected the Taylor influence in his *Essays on Agriculture,* which was published in 1819. When compared with Taylor's views, the kinship of ideas and emphasis may be seen in the following so-called "First Principles of Agriculture," as set forth in Jeffreys' volume.

> Industry and attention to agricultural pursuits, and intelligence therein, are indispensable to insure success.
>
> . . . experienced and skilful farmers have in all ages discovered the necessity and utility of draining wet lands. . . .
>
> One of the most important principles in agriculture is cleanly farming. . . . Every field in cultivation should be kept entirely clean.
>
> Manures will always fail in producing the desired effect, in proportion as draining and cleaning are neglected.
>
> A change, or a judicious rotation of crops is necessary, in order to keep the soil in good heart and to enable it to produce its utmost.
>
> Selecting and propagating . . . the most approved kinds of grain and seeds, is the surest method of preserving them in perfection. Seeds should be selected in the same manner that breeders are selected.
>
> Liberality in procuring good tools or implements for the hands on a farm is the economy of agriculture.
>
> Foresight is another item in the economy of agriculture. It consists in preparing work for all weather, and doing all work in proper weather, and at proper times.
>
> When a thing is done, let it be well done, and it will not require to be done soon again. . . .[65]

[63] *Ibid.,* 704.

[64] *Farmer's Register,* VIII (December, 1840), 704.

[65] (George W. Jeffreys), *Essays on Agriculture,* 6-15, *passim.* Compare with John Taylor's remarks on "The Economy of Agriculture," reprinted in *Farmer's Register,* VIII (December, 1840), 755-757.

Neither Taylor nor Jeffreys understood the chemistry of soils, yet they both put great emphasis on a generous use of manures in planting. The farmer, wrote Taylor, could no more expect to reap good crops by neglecting his manures than he could expect Heaven by neglecting his morals.[66] Both urged the use of vegetable as well as animal manures. Jeffreys, like Taylor, highly praised corn as an all-purpose crop for the farmer. It provided food for both man and beast and an abundance of litter for use as a manure. The litter of all crops, leaves from the trees, scrapings from fence corners and ditch banks, and muck from lowlands and swamps were recommended by both for use as manures.

Nothing sensational resulted from these proposals; perhaps the great majority of farmers did not even hear of them, yet evidences began to appear that agriculture in North Carolina was on the threshold of a new day. One evidence of this new spirit is seen in an article by a farmer from Newbern published in the *American Farmer*. "The farmers," this correspondent wrote, "in this vicinity are turning their attention to lime as a manure. We have an abundance of the stone called *shell-rock*—it yields excellent lime, but is very hard and a portion is frequently left unchanged."[67] The author requested information on building kilns and burning lime.

In North Carolina, as in other places, many obstacles stood athwart the path of progress. This was particularly true in agriculture, the so-called "oldest of the arts." In the absence of free public schools, illiteracy rates stood high in the state and science and learning were held in ill repute by the mass of people. The state's reputation was bad in most respects. William R. Davie wrote in 1805 that "the friends of science in the other States regard the people of North Carolina as a sort of semi-barbarians, among whom neither learning, virtue nor men of science possess any estimation."[68] Archibald DeBow Murphey's remarks upon this score were almost as severe. In 1819 he wrote Thomas Ruffin as follows:

> I know ten times as much of the Topography of this Circuit, as the Men who have lived here fifty years. I had no idea that we had such a poor, ignorant, squalid Population, as I have seen. . . . In the Towns are found decent and well informed Men in Matters of Business, men who look well and live well. But the Mass of the Common people in the Country are lazy, sickly, poor, dirty and ignorant.[69]

[66] *Farmer's Register*, VIII (December, 1840), 723; (George W. Jeffreys), *Essays on Agriculture*, 30-56. For an excellent summary of Taylor's contributions to agricultural reform, see A. O. Craven, *Soil Exhaustion*, 102-103.

[67] *American Farmer*, II (July, 1820), 136.

[68] Quoted in Kemp P. Battle, *History of the University of North Carolina*, I, 178.

[69] W. H. Hoyt (ed.), *Papers of Archibald D. Murphey*, I, 138. R. D. W. Connor has described Murphey as "The keenest-minded and most far-sighted statesman that North Carolina has yet produced. . . ." R. D. W. Connor, *Rebuilding An Ancient Commonwealth*, I, 19-20. Professors Lefler and Newsome identified

Even the intelligent and "improving" farmer labored under severe handicaps. To a great extent he was unacquainted with the principles of plant nutrition, the chemistry of soils, and how tillage helps plants to grow. As a result of his ignorance of the soils, his efforts at fertilizing them were unscientific and often resulted in failure of crops and injury to the soil. He had no agricultural schools or experiment stations to aid him in solving his problems. His implements were but little advanced in durability and efficiency from those used in medieval times. His farm animals were in the main of a "common" breed. But he was just entering on the threshold of an era that witnessed momentous efforts, to which North Carolinians made important contributions, to correct and improve the agricultural situation. Being by nature conservative and cautious, the farmer was skeptical of innovations and experimentation. This attitude tended to retard severely the progress of agriculture in the state. "In addition he was bound by a superstitious dependence upon the phases of the moon as a factor in farm procedure; this was generally accepted as agricultural science."[70]

In spite of the work, advice, and examples of such contemporary leaders as Taylor, Jeffreys, and the number of unknown "improving" farmers in nearly every community, the pioneer system of agriculture obtained in North Carolina until well into the nineteenth century. As noted above, the system imposed on the land an excessive burden in the production process and ultimately resulted in impoverishment of the soil. Charles Fisher described the system then in use to the Rowan Agricultural Society, in 1821, as follows:

> We pursue a course of agriculture that takes all from the earth, and returns nothing to it: We go on, year after year, tilling our fields, without any pains to return to the earth the strength that each crop takes from it. We completely exhaust our soil by an unvaried succession of crops; and, when it can produce no longer, we turn it out into fields, let it wash into gullies, and grow up with pines, and broom sedge, that never failing symptom of exhaustion. This is the common fate of our fields; the system that is defacing our country, and ruining our lands.[71]

---

Murphey as "an intelligent, well-educated, public-spirited" man who was the "genius, leader, and mouthpiece of the progressive reform movement in North Carolina after 1815. . . ." H. T. Lefler and A. R. Newsome, *North Carolina,* 313.

[70] A. L. Demaree, *American Agricultural Press,* 5.

[71] Quoted in *American Farmer,* III (August, 1821), 169. For Davie's description of farming methods and wastage of the soil, see *American Farmer,* I (October, 1819), 218-219. Jeffreys pointed out the causes of loss of soil fertility in his *Essays on Agriculture,* 16-17, 21-24. As a remedy, he proposed: "Let them be recruited by deep and horizontal ploughing, a favorable rotation of crops, and all the manures which the most skillful and industrious management will afford." *Ibid.,* 24. For the more recent explanations of soil exhaustion in the South Atlantic States in this period, see A. O. Craven, *Soil Exhaustion,* 11-24, 36-38, 83; L. C. Gray, *History of Agriculture,* I, 448; II, 910-911; R. H. Taylor, *Slaveholding in North Carolina,* 43.

Loss of soil fertility and the growing scarcity of good lands in the older southern states were among the principal causes of the continued and increasing flow of emigration from those states. That flow was increasing in North Carolina in the 1820's. William R. Davie pointed out the connection between soil exhaustion and emigration in a speech to the South Carolina Agricultural Society, in 1819.

> It is a lamentable fact, that the generality of our citizens cultivate their lands, as if they were but tenants at will—as if there was a general presentiment, that we were all sooner or later destined to join in the current of western emigration; and upon our present wretched system of agriculture, this presentiment is not illy founded. Apathy, or an ill-directed avarice, looking only to the advantages of the present moment, may operate as decisively as the mandate of a despot to banish us from our country.[72]

The urge to leave the state affected all ranks of society and in some communities became almost an epidemic. Some left North Carolina without even bothering to dispose of their lands.[73] With most, however, it was a more orderly process. Lands and other property not intended for the new home in the Lower South or the West were put on the block. The constancy of such actions had the effect of keeping the price of land deflated.[74] Capital, which was already scarce, was constantly drained from the state as citizens liquidated their estates and moved away. The Salem Diary noted such a movement on October 14, 1796.

> For some weeks almost daily one or more families have camped in our neighborhood on their way to Kentucky or Cumberland. . . . In today's camp there were over one hundred and twenty persons, old and young, with twelve four-horse wagons, more than twenty other horses, and over thirty head of cattle.[75]

Archibald D. Murphey estimated that between 1790 and 1816 the state lost more than 200,000 people by emigration.[76] His explanation for this is to be found in the following remarks:

[72] *American Farmer,* I (October, 1819), 218. In a committee report to the state legistature in 1828, Fisher spoke of the state's condition as one of "embarrassment never before equalled in North Carolina. . . ." Because of the depressed prices of cotton, tobacco, rice, and naval stores, he continued ". . . at this time, they are scarcely worth producing. . . . Tobacco is now taxed in the British markets, more than 600 per cent while the demand for cotton, our other great staple, does not keep pace with its increased production." Reproduced in *Report on the Geology of North-Carolina. . . .*, Part III, 46-47.

[73] J. G. deR. Hamilton (ed.), *The Papers of Thomas Ruffin,* I, 216. For an account of the state's losses resulting from emigration, see James William Williams, "Emigration From North Carolina: 1789-1860," 85-93. Unpublished masters thesis at the University of North Carolina, 1939.

[74] A. R. Newsome (ed.), "Twelve North Carolina Counties in 1810-1811," *N. C. H. R.,* VI (April, July, 1929), *passim.*

[75] Adelaide L. Fries (ed.), *Records of the Moravians,* VI, 2564.

[76] W. H. Hoyt (ed.), *Papers of Archibald D. Murphey,* II, 20.

North-Carolina has not, since the year 1784, cultivated a respect for herself, nor has she sought to acquire a character in the Union. She has remained careless and spiritless, until her citizens have formed not only political, but commercial and local feelings with the people of Virginia and South Carolina. . . .

A conviction of the reproach which we suffer, and the want of encouragement to industry and enterprise, have driven from the State, a large portion of her most useful population.

. . . the true cause is to be found in the want of public spirit, of State pride, and of State feeling.[77]

James Dunwoody Brownson DeBow observed that the loss to the state was greater than a mere loss of population and capital. " . . . the moral and social condition," he wrote, "of those who remain, as well as those who remove, must ever be disturbed, and more or less retarded in their advancement."[78]

It is impossible to appraise the complete loss to the state, whether of capital or population; but in the number who left North Carolina were three young men who were later to become President of the United states.[79] To understand the motivating and driving forces behind this movement and to counteract them, in so far as possible, became the goal of North Carolina's economic, social, and political leaders alike, in the ante-bellum period. More will be said on this problem later.

Most contemporary accounts agree that economically, politically, and socially, North Carolina was in a serious crisis in 1820. "In North Carolina," wrote Jeffreys, "the state of agriculture is at the lowest ebb; I speak not this with reproach, but with the deepest regret. . . . Our present, is a land-killing system, which must be altered for the better; for if persevered in, it must ultimately issue in want, misery and depopulation."[80]

The end of the War of 1812 was followed by a change of policy in most European states regarding importation of food crops, with the result that the high prices for farm products enjoyed before and immediately after the war soon collapsed. The Panic of 1819 contributed to lessen the demands of the domestic market and to drive agricultural prices further downward. A correspondent to the *American Farmer* from Albright, North Carolina, wrote as follows in 1820:

Things have undergone a great revolution here as in other places . . . some are groaning under the burthen which better times brought on them, and resolving that if they can extricate themselves now, that nothing shall ever involve them again. Money is very scarce indeed. Wheat is worth but 50 cents—Corn $1 to $1.25 per barrel—Brandy 50 to 60 cents—whiskey 37½ cents, but will soon be lower—Cotton and Tobacco—not selling—Pork, $3.50

[77] W. H. Hoyt (ed.), *Papers of Archibald D. Murphey,* II, 177-178.
[78] J. D. B. DeBow, *Industrial Resources,* II, 109.
[79] Andrew Jackson, James K. Polk, and Andrew Johnson.
[80] (George W. Jeffreys), *Essays on Agriculture,* 5.

to $4—Beef, 2½ to 3 cents per lb—Labour still keeps up—very far above the depreciation in other things.[81]

Relatively, North Carolina in 1820 stood low, in comparison with other states, in the measure of progress made since 1783. Many explanations were offered by contemporary and later writers of the state's lowly status.[82] A correspondent of a state newspaper in 1825 expressed a view, held by many thinking men in the state, when he wrote:

It is humiliation in the highest degree to behold the gigantic strides by which our sister states have surpassed us in the march of improvement. . . . Our agriculture is nearly what it was in the days of our fathers; enterprise, of every kind, seems to have taken wings and fled to some congenial abode. . . . It is now high time to retrieve our lost honor, and establish our character for intelligence, patriotism and enterprise.[83]

Despite the dark picture painted by the writer, significant beginnings had been made in drawing attention to the state's agricultural deficiencies and the ways to their correction, which, during the ante-bellum period, were to improve the standing of the state. The story of that progress will be told in later chapters.

[81] *American Farmer,* II (December, 1820), 311.

[82] Connor's appraisal and conclusions drawn from the various contemporary writers is excellent. R. D. W. Connor, *Rebuilding an Ancient Commonwealth,* I, 466, 469-474.

[83] Salisbury *Western Carolinian,* January 11, 1825.

CHAPTER III

# LAND HOLDING AND LABOR

In differentiating between the plantation and the farm, Ulrich Bonnell Phillips has observed that the plantation was particularly characterized by a large labor force, usually slaves, who worked in routine manner, under supervision, with the primary purpose of producing a staple product for sale.[1] The farm, on the other hand, had a relatively small labor force with no sharp distinction between workmen and supervisor, where no regular routine was followed, and the primary purpose was divided between producing for home use and outside sale. The farmer might or might not be a slaveowner or an employer of hired labor.[2] Many different definitions have been used for these terms, but the above will be followed in this study.[3]

Whether a unit could be developed as a farm or as a plantation was determined, to a considerable extent, by the size of the land holding. To begin with, the size of landholdings in North Carolina in 1783 were approximately as follows: one-fourth were from 1 to 100 acres; one-fourth were from 101 to 200 acres; one-fourth were from 201 to 400 acres; and a little over one-fourth were of 401 acres each or above.[4]

Between 1783 and 1860 radical changes were made in those percentages and a considerable reduction in acres of the average holding took place. In 1860, the number of owners having acreages in excess of 100 represented only about thirty per cent of the whole. The number owning in excess of 500 acres made up only about two per cent of the total. And, only 311 persons owned 1,000 acres or over.[5] The average number of acres per farm in 1860 was 316, having fallen in ten years from 369. On the other hand, the number of farms increased from

[1] U. B. Phillips (ed.), *Plantation and Frontier Documents: 1649-1863,* I, 72.
[2] *Ibid.,* I, 72-73.
[3] L. C. Gray, *History of Agriculture,* I, vii-viii; E. Q. Hawk, *Economic History of the South,* 234. For an excellent discussion of differences between farms and plantations, see Robert R. Russel, "The Effects of Slavery upon Nonslaveholders in the Ante Bellum South," *Agricultural History,* XV (April, 1941), 114-120.
[4] F. G. Morris and P. Morris, "Economic Conditions in North Carolina about 1780," *N. C. H. R.,* XVI (April, 1939), 128-129.
[5] *U. S. Census of 1860* (Agriculture), 210. At the same time, South Carolina, with only three-fifths as much land area, had 482 owners of 1,000 acres or more. Virginia, with a somewhat larger area suited to agriculture, had over twice as many owners in this category. *Ibid.,* 221. Twenty-one of North Carolina's eighty-six counties did not contain a single holding of over 1,000 acres. Fifteen of those were western counties. *Ibid.,* 210.

56,963 in 1850 to 75,203 in 1860.[6] Unquestionably the farm as a producing unit, rather than the plantation, was predominant in North Carolina agriculture in 1860.[7]

Several factors combined to encourage a broad participation in land ownership in North Carolina between 1783 and 1860. The urge to acquire land, which characterized colonial times, continued to be the "aim of all action" for the great majority of white citizens of the state. Agriculture, as in the colonial period, continued to be the principal economic interest of an overwhelming proportion of the population. Because of emigration from the state, which continued throughout the period and sometimes threatened depopulation of whole communities, the price of land was constantly depressed below its real value.[8] Opportunities to buy land, therefore, were frequent and the terms of purchase easy. In spite of the population loss to other states, the number of white people in North Carolina increased over 100 per cent between 1783 and 1860. The tendency towards engrossment by planters of the lands suited to the staples was more restricted in North Carolina because of the state's isolation from good markets, and the limitations imposed by geography.

From the above, it may be seen that the trend in land ownership in most of North Carolina in the ante-bellum period was definitely contrary to the traditional concept of the ever-expanding plantation. Actually the average producing unit was growing smaller, and, in most, the primary purpose of production was self-sufficiency. North Carolina, because of her tardiness in acquiring adequate transportation and marketing facilities and her long experience in self-sufficing farming, had perhaps proportionately fewer units which concentrated entirely on the production of a staple crop than any other slave state.

Houses, barns, outhouses, fences, and grounds of the average ante-bellum North Carolina farm or plantation were but little improved over those of the colonial period. To outsiders, deficiencies in these respects were particularly noticeable. In 1816, Lieutenant Francis Hall, on a trip through North Carolina from Richmond to Charleston, observed that "The houses are all built of scantling, and are worse than any thing

[6] *Ibid.*, 222. The average size of farms in Virginia and South Carolina in 1860 was 324 and 488 acres respectively. *Ibid.*

[7] The census of 1860 credited North Carolina with 121 planters and 85,198 farmers. *U. S. Census of 1860* (Population), 362-363.

[8] Henry King Burgwyn, one of the most widely traveled, intelligent, and progressive planters in the state, speaking of the lands along the Roanoke River where he lived, said: "from my extensive experience, I am convinced that nowhere in the eastern portion of the U. States are lands so low in price in proportion to their real value . . . these lands in Virg[ini]a and Maryland are worth from 75 to $100 per acre, here but little over 1/3d that sum. . . . ." J. G. deR. Hamilton (ed.), *Papers of Thomas Ruffin*, II, 577-579. The lands about which this statement was made were located in one of the wealthiest farming regions of the state, which, at that time, had communication with a splendid market at Norfolk by both water and rail!

in the form of dwellings, but the negro huts; for they are penetrable at every crevice; while, from the usual mildness of the weather, doors have become altogether released from the duty of being shut."[9] Another British visitor in Buncombe County in 1843 was even less favorably impressed. ". . . though there were some few decent houses," he wrote, "the greatest number of them were miserable-looking dwellings, rude and dirty without, and bare and comfortless within; the broken panes of glass being more numerous than those that were whole, and everything bespeaking the indolence and dirty habits of the occupants."[10] A native reporter observed that most houses were of either log or frame construction, with brick chimneys, and described them as being "comfortable but seldom eligant."[11]

The wealthier planters, of course, were able to build more for permanence and comfort. Thomas P. Burgwyn not only had a large brick barn in which he threshed his wheat by steam, but surrounded his house with a garden especially designed and laid out by an English gardener. In the center was a fountain for irrigation purposes; and, in addition to its ornamental aspects, this garden was used to produce the plantation supply of vegetables. Burgwyn maintained attractive, clean, and comfortable houses for his slaves, "a perfect town of them," in which certain buildings were equipped as a nursery, chapel, and a hospital.[12] His brother Henry K. Burgwyn made equally as good provision for his comfort and that of his slaves. A Northerner who visited his plantation for a week in 1855 said he "provides so well for the animal happiness of the slave that it necessitates one to continually summon up his principle to resist falling in with it and heartily approving the whole system."[13]

The most intelligent and, perhaps, extensive examples of improvements of the homestead were those made by Josiah Collins and Ebenezer Pettigrew, who occupied adjoining plantations on Lake Phelps in Tyrrell County. The mere occupation of their lands contemplated an intricate and expensive system of drainage, dyking, and reclamation of heavily forested swamp lands, which they and their forebears accomplished over a long period of years. Both Collins and Pettigrew, by means of a network of canals controlled by flood gates, could harvest their entire crops into flat boats for easy transport to their four-story barns where the harvest would be lifted out and stored by water driven

[9] Francis Hall, *Travels in Canada and the United States, in 1816 and 1817,* 241. Frederick Law Olmsted visited the state over the same route in 1853 and was equally as severe in his criticism of North Carolina houses. *A Journey in the Seaboard Slave States With Remarks on Their Economy,* 321, cited hereinafter as *Journey in the Seaboard Slave States.*

[10] James Silk Buckingham, *The Slave States of America,* II, 198.

[11] A. R. Newsome (ed.), "Twelve North Carolina Counties in 1810-1811," *N. C. H. R.,* VI (April, 1929), 178.

[12] *American Farmer,* VI (March, 1851), 307-308.

[13] Quoted in U. B. Phillips, *Life and Labor in the Old South,* 253-255.

elevating machinery. Each operated a flour mill, grist mill, saw mill, threshers, shellers, fanners, and smaller implements such as grindstones, by the water power which he had harnessed.[14]

As guides for the general improvement of buildings and grounds, farm papers, toward the end of the period began to carry articles and cuts on the proper layout and development of the home site. Diagrams of floor plans and instructions on the location of the various buildings of the model homestead began to assume a prominent place in such publications. Such features too as slave quarters, cellars, barns, sheds, arbors, trellises, ice houses, carriage houses, stables, granaries, tool houses, work shops, lightning rods, etc., began to receive conscious planning.[15] State agricultural leaders threw their influence behind the movement for improvement in this respect, Paul C. Cameron, leading planter of Orange County, urged members of the county agricultural society "seek to make your houses lovely. . . . No cottage so humble that it may not be made sweet by neatness; none so lowly that it may not be embellished by the hand of industry."[16]

Boundaries between land holdings were often marked by rail fences. It was compulsory that cultivated fields be fenced because the "open range" prevailed throughout the period of this study.[17] These facts combined to give fence building and repairing a prominent place in the annual work program of nearly every agricultural unit, and to make the rail fence a feature of the ante-bellum landscape. A well kept and adequate fence, then, as now, was accepted as a proper reflection of enlightened self-interest by the proprietor.

The self-supporting, "worm" type rail fence was most commonly used and thought to be cheapest where timber abounded. "The timber is an encumbrance," wrote a contemporary, "and therefore cost[s] nothing; and the rails can be cut and split for 50 to 75 cents per 100, and the hauling and placing is still less."[18] Prodigious numbers of rails were used. Richard Mitchell spoke of having used 30,000 in fencing the major part of his 350 acre plantation near Newbern.[19] Writing of his

[14] *Farmer's Register,* VII (November, 1839), 702-703; (December, 1839), 725-733; Bennett Harrison Wall, "Ebenezer Pettigrew, An Economic Study of An Ante-Bellum Planter," *passim.* Unpublished doctoral dissertation at the University of North Carolina, 1947, cited hereinafter as "Ebenezer Pettigrew."

[15] Richard Lamb Allen, *The American Farm Book; or Compend of American Agriculture. . . . ,* 302-319, cited hereinafter as *American Farm Book.*

[16] Quoted in *Arator,* I (June, 1855), 73.

[17] The legislature in 1777, in the interest of "peace and harmony of every neighborhood," prescribed that "every planter shall make a sufficient fence about his cleared ground under cultivation, at least five feet high. . . ." *Laws of N. C.* (rev.), I, 344. In 1791, another act made failure to maintain proper fences around growing crops an indictable offense, and prohibited anyone from killing livestock, belonging to another, that happened to enter upon improperly fenced land. *Ibid.,* 677-678.

[18] R. L. Allen, *American Farm Book,* 284-285.

[19] Edenton *State Gazette of North Carolina,* November 15, 1792.

visit on Henry K. Burgwyn's plantation in 1849, Solon Robinson said: "Mr. B. told me that there are about thirty miles of fencing upon these places [his and his brother's], *to keep out other folks cattle.* What a tax! But it is just so all over the United States."[20] Fence building and maintenance, therefore, was a necessary but expensive and labor-consuming activity of nearly every land owner.[21]

The search for cheaper and more effective fencing early attracted the attention of agricultural reformers. Experimentation was directed principally toward finding a suitable live fence.[22] John Taylor of Caroline recommended the use of cedar trees for that purpose.[23] William J. Bingham, of Alamance County, experimented with the osage orange.[24] Charles Fisher recommended the use of cedar and white thorn in Rowan County. Such a fence would save much lumber and labor, insure the safety of crops, and add to the appearance of the farm, he wrote.[25] A contributor to the *Star* cautioned that a live fence must be strong to keep cattle out because they were used to forcing their way through brambles, briars, and obstructions of all sorts.[26]

The proper organization and direction of labor on the farms and plantations of North Carolina were two of the most important problems of the land owners. With the non-slave-owning farmer, in 1860 about seventy per cent of the white families were in that group, the solution was easy. He simply did all the work himself with the assistance of his family, and, perhaps, an occasional hired hand.[27] With the slaveowner, however, these problems became more complex as the slaveholding became larger. Conditions affecting slave distribution, prices, housing, subsistence, efficiency as workers, and management were of great concern to slaveowners.

As noted previously, at the time of the first census the bulk of the slave population was located in the rice and tobacco producing areas of the Coastal Plain. With the expansion of tobacco toward the northern Piedmont and the development of cotton culture in the lower Piedmont, slaveholding became more general in the state. This trend toward

[20] Herbert Anthony Kellar (ed.), *Solon Robinson, Pioneer and Agriculturist,* II, 230. Cited hereinafter as *Solon Robinson.*

[21] Fences were usually built in the winter at the time the new grounds were being cleared of trees. Fence repairing was usually done in the fall and winter in connection with the job of gathering the annual supply of firewood. See J. G. deR. Hamilton (ed.), *The Papers of Thomas Ruffin, passim;* Stephen Andrews Norfleet Papers, *passim,* Southern Historical Collection, Library of the University of North Carolina, hereinafter cited as Norfleet Papers.

[22] For a list of various plants suggested for use as live fences, see R. L. Allen, *American Farm Book,* 283-288.

[23] Raleigh *Star,* February 22, 1810.

[24] J. G. deR. Hamilton (ed.), *The Papers of Thomas Ruffin,* II, 428-429.

[25] *American Farmer,* III (August, 1821), 170.

[26] Raleigh *Star,* January 4, 1810.

[27] Although the primary aim of production of the farm was self-sufficiency, the farmer could produce cotton and tobacco just as economically as the large scale planter.

diffusion of slave population and ownership continued throughout the ante-bellum period. The following table illustrates somewhat the direction and extent of such changes in the last decade of the period.[28]

Slaveholding in North Carolina, 1850-1860[29]

| Number of Slaves | Number of Owners, 1850 | Number of Owners, 1860 |
|---|---|---|
| 1 | 1,204 | 6,440 |
| 2 and under 5 | 9,668 | 9,631 |
| 5 " " 10 | 8,129 | 8,449 |
| 10 " " 20 | 5,898 | 6,073 |
| 20 " " 50 | 2,828 | 3,321 |
| 50 " " 100 | 485 | 611 |
| 100 " " 200 | 76 | 118 |
| 200 " " 300 | 12 | 11 |
| 300 " " 500 | 3 | 4 |
| Total number of owners | 28,303 | 34,658 |
| Total slaves | 288,548 | 331,059 |
| Average holding | 10.1 | 9.5 |

An examination of the census returns for 1860 reveals that eighty-five of the eighty-six counties, into which the state was divided at that time, had at least one owner who had as many as thirty slaves. Eighty-one counties had owners of as many as thirty-nine slaves. Seventy had owners of as many as forty-nine slaves. Fifty-one had owners of as many as one hundred slaves. Only one hundred and thirty-three owners had in excess of one hundred slaves. Of this number, twenty-four were evenly divided between Edgecombe and Halifax, two counties of the Coastal Plain which showed strong tendencies in the later ante-bellum period to develop the traditional plantation-type economy. It is significant to note that between 1850 and 1860 the number of owners holding just one slave increased by 5,236, while the owners of two or more slaves increased only by 1,119. Unquestionably, as the period drew to a close, slaveowning was becoming more popular in the farmer class.[30]

Generally, the largest slaveholdings were located in the East. The average for Edgecombe County was 15.4; for Bertie, 17.4; for Northampton, 12.5; and for Halifax, 14.8. All of these were leading cotton producing counties in 1860. The largest holdings in the West were in the tobacco producing counties of Granville, Caswell, and Warren, where the average holdings were 11.2, 12.3, and 20.2 respectively.[31]

[28] *U. S. Census of 1860* (Agriculture), 247-248. For further tables and discussion on this subject, see *ibid.* (Population), 598-604; L. C. Gray, *History of Agriculture,* I, 530 532; J. S. Bassett, *Slavery in the State of North Carolina,* 78-79; R. H. Taylor, *Slaveholding in North Carolina,* 17-19, 37, 45, 65.

[29] *U. S. Census of 1860* (Agriculture), 235-236.

[30] *U. S. Census of 1860* (Agriculture), 235-236.

[31] R. H. Taylor, *Slaveholding in North Carolina,* 35-36. For an account of the distribution of slaves by regions within the state in 1860, see Joseph Carlyle

After the introduction of cotton, the price of slaves rose slowly until the late forties and then, as more prosperous times came, sky-rocketed to an all-time high in the eighteen fifties. John Haywood wrote from Raleigh in 1808: "Negroes sell at what I call a very high price:—likely men from 400 to 500 Dollars—stout Boys from 350 to 400 Dollars—and Women from 300 to 375 Dollars."[32] He observed, however, "Colo. Polk now and then picks up one . . . on much more advantageous terms: but he has the advantage of being applied to by the Sellers, as a monied Man. . . ."[33] Able-bodied males between the ages of sixteen and thirty were in greatest demand. Men with skills, however, commanded the highest prices. In 1854, in Edgecombe County "likely" Negro men were selling at from $1,000 to $1,200 and a twenty-three year old rough carpenter brought $2,000. Young women and girls ranged in price from $700 to $900.[34]

Though North Carolina lost many Negroes, as well as whites, to the southern and western migration and is usually classified as a "selling" state, slavery had a much stronger hold on the community at the end of the period than in the beginning. Between 1790 and 1860 the ratio of slave population to the whole population in the state rose from 25.5 to 33.4.[35]

Many planters and farmers added to their labor supply by hiring slaves, usually the property of minor children or of unsettled estates. This was an annual event in every county and was supervised by the county courts.[36] A correspondent of the Goldsboro *Patriot* wrote: "Much animation was exhibited at the negro hiring in this place on Saturday last . . . prices for men ranged from $125 for common to $200 for extra. . . . These prices are higher than have ever been known heretofore in this section of the country."[37]

Most operators wishing to increase their labor supply preferred to hire slaves rather than free Negroes or white laborers. Calvin Jones, of Wake Forest, expressed that preference in 1828 as follows: "In general, white labor, for agricultural purposes, is not desired. It is more expensive in the hire; more troublesome and costly in the maintenance; and much less steady and efficient in the application."[38] Hinton Rowan

---

Sitterson, "Economic Sectionalism in Ante-Bellum North Carolina," *N. C. H. R.*, XVI (April, 1939), 135-137, 139-141, 144.

[32] H. M. Wagstaff (ed.), *Papers of John Steele*, II, 533.

[33] *Ibid.*, II, 535.

[34] Tarboro *Southerner*, January 14, 1854. The trend of slave prices in the fifties was steadily upward in the South Carolina, Virginia, Georgia, and Louisiana markets, rising from a general average of $680 to approximately $1,335 for prime field hands. See chart of these price changes in U. B. Phillips, *American Negro Slavery*, 370. See also, R. H. Taylor, *Slaveholding in North Carolina*, 69-74.

[35] L. C. Gray, *History of Agriculture*, II, 651-656.

[36] For an excellent discussion of the various aspects of this subject, see R. H. Taylor, *Slaveholding in North Carolina*, 74-79.

[37] Quoted in Raleigh *Register*, January 8, 1853.

[38] *American Farmer*, X (November, 1828), 273.

Helper, writing nearer the end of the period, did not agree that slaves were more efficient or less expensive than white laborers. He said slaveholding had become so popular that non-slaveowners were being carried away by the fashion "to have Negroes around," and were actually willing to pay more to hire a slave than to employ a "sober, energetic white man."[39]

The cost of upkeep of a slave varied, of course, with time and place. In a committee report to the House of Commons in 1828, Charles Fisher of Rowan County said: "Some very intelligent gentlemen, large owners of slaves, give it as their opinion, that slaves . . . can be decently clothed and fed for 25 dollars each . . . per year. . . ."[40] In the main, each slaveowner sought to produce on his holding enough subsistence crops to provide for his family and labor force, and, to a great extent the slave's clothing, including shoes, was made on the plantation.[41] Many owners permitted their slaves to plant patches of their own and to own livestock. This privilege was subject to the owner's control as to disposal of any surpluses resulting.[42]

Under the guiding discipline of plantation organization the slave early demonstrated an aptitude for craftsmanship and developed in efficiency as a worker. All slaves, however, had neither the aptitude nor the opportunity to develop skills, but the tendency of the slave system was to train selected workers so that the producing unit would be as nearly self-sufficing in its labor requirements as possible. Charles Fisher made the following unchallenged statement in an important committee report to the North Carolina House of Commons in 1828: "What branch of mechanics have we in our country, in which we do not find negroes often distinguished for their skill and ingenuity? In every place we see them equalling the best white mechanics."[43] Stephen A. Norfleet noted the death of one of his skilled workers as follows: "Con-

[39] Hinton Rowan Helper, *The Impending Crisis of the South: How to Meet It,* 380. Cited hereinafter as *Impending Crisis.* Helper made a study of wages paid free white labor and for the hire of slave labor, for agricultural purposes, in his community and found that white laborers were employed for an average of $84 per annum—including board only; whereas, an average of $115 per annum was being paid for the hire of slaves—including board, clothing and medical attendance. Thomas L. Clingman, a representative from North Carolina in the United States Congress, said in a speech to that body with reference to labor conditions in the state that ". . . either free or slave labor can be obtained at forty to fifty cents per day." Quoted in J. D. B. DeBow, *Industrial Resources,* II, 177.

[40] Quoted in *American Farmer,* IX (January, 1828), 353. See also *ibid.,* X (November, 1828), 273; R. H. Taylor, *Slaveholding in North Carolina,* 92-93.

[41] The diary of operations maintained by Stephen A. Norfleet for his plantation, "Woodburn," located in Bertie County, indicates that it was almost routine practice that slave women remain inside during inclement weather to work on clothing. He hired some of the sewing done. A frequent entry was: "One man making shoes." Norfleet Diary, *passim,* Norfleet Papers.

[42] Norfleet Diary, *passim,* Norfleet Papers; R. H. Taylor, *Slaveholding in North Carolina,* 89-92.

[43] Quoted in *American Farmer,* IX (January, 1828), 353.

nor died very suddenly about 11 o'clock P.M. He will be very much missed as he was shoemaker, tanner, cooper, Distiller and useful generally . . . his loss will be very sensibly felt."[44]

Many skilled slaves hired their own time in order to practice their trade. Toward the end of the period, however, as the white mechanics became more articulate, serious objections were raised to this practice.[45] In 1850, the white mechanics of Washington, North Carolina, formed an organization and resolved that henceforth they would not employ any Negro mechanic "or learn any negro boy a trade." They condemned the practice of hiring a slave his time.[46]

Toward the end of the period many slaves were employed in mines, the turpentine industry, small factories, public works, and as personal servants, but the great majority of them, as was also the case with white people, was engaged in agriculture. Where properly directed and provided for, the slave grew in efficiency as a laborer and in loyalty to his owner. Where improperly supervised and mistreated, he rewarded his master with work carelessly and slovenly performed and often by running away.[47] The search for a solution to the problems of slave management continued throughout the period with agricultural reformers, and with large and small owners alike. A correspondent of *De Bow's Review* wrote that one good article on the subject of slave management "would be worth more to the master than a hundred theories about 'rotations' and 'scientific culture'; and infinitely more to the slave than whole volumes dictated by a spurious philanthropy looking to his emancipation."[48]

Frederick Law Olmsted observed that in some respects the condition of the slave in North Carolina was "less lamentable" than in other parts of the South and the institution had "more of patriarchal character." "The slave more frequently appears," he wrote, "as a family servant—a member of his master's family, interested with him in his fortune, good or bad . . . slavery thus loses much of its inhumanity."[49]

[44] Norfleet Diary, Norfleet Papers.

[45] R. H. Taylor, *Slaveholding in North Carolina,* 75.

[46] Tarboro *Press,* August 31, 1850. The name of this publication changed several times but is cited as Tarboro *Press.*

[47] Practically every issue of every paper published in the period carried advertisements for one or more runaway slaves. Some made good their escape from slavery, but most, perhaps, were retaken after spending some time lurking in the neighborhood. Many sought refuge in such inaccessible places as the Dismal Swamp. Because these runaways were "corrupting and seducing other slaves" and "greatly impairing the value of slaves in the district," the state legislature, in 1822 and in 1847, passed laws designed to encourage their recapture. *Laws of N. C.* (1822), 28-29; *Laws of N. C.* (1846-1847), 109-113. For an excellent account of troubles experienced by one North Carolina planter with runaways, see B. H. Wall, "Ebenezer Pettigrew," 124-134.

[48] Quoted in J. D. B. DeBow, *Industrial Resources,* II, 330. For further remarks on the relationships between slaves and owners, see *ibid.,* 330-338; *Farmer's Journal,* II (May, 1853), 53-54.

[49] F. L. Olmsted, *Journey in the Seaboard Slave States,* 367.

Olmsted explained the more satisfactory relations between whites and slaves in North Carolina by quoting a slave from the state whom he met in South Carolina as follows: ". . . up in our country . . . dey doan' hev no real big plantation, like dey has heah, but some on 'em has ten or twelve niggars, may be, and dey juss lives and talks along wid 'em; and dey treats 'um most as if dem was dar own chile. Dey doan' keep no niggars dey can't treat so. . . ."[50]

Hinton Rowan Helper, who reflected to a considerable extent the anti-slavery sentiments of non-slaveowning whites in the South, also thought the slaveowners in North Carolina were "more moderate, decent, sensible, and honorable, than the slaveowners in either of the adjoining states, or the states further South. . . ."[51]

Thomas Ruffin, leading jurist, a planter who owned over one hundred slaves, and several times president of the North Carolina Agricultural Society, perhaps expressed the prevailing sentiment of the slaveowners of the state on the subject of slave treatment when he wrote:

> Slavery, indeed, is not a pure and unmixed good. Nor is anything that is human. There are instances of cruel and devilish masters, and of turbulent and refractory slaves, who cannot be controlled and brought into subjection but by extraordinary severity. But these are exceptions, and rare exceptions. Great severity in masters is as much opposed to the usuages [*sic*] of our people as to the sentiment of the age, and, indeed, to the interest of the master.[52]

Though undoubtedly, the general aspects of slavery in North Carolina were less objectionable than in most southern states, the attacks made on the institution called forth from North Carolinians reaction similar to that elsewhere in the South. Prior to about 1830, public opinion in the state on slavery was unimpassioned, but after that date a combination of circumstances tended to make the conditions of slavery more severe, and to develop attitudes of intolerance of criticism of the institution. These circumstances grew out of such things as fear of slave insurrections; the rise of abolitionism in the North; the growth of free Negro population in the state; animosity of the white artisan class to Negro competition; contests over extension of slavery into the territories; and the publication of Helper's *Impending Crisis*.[53]

Most farmers and some planters who owned only a small number of slaves acted as their own overseers, often working side by side with their slaves. Ebenezer Pettigrew was one of these throughout a great part of his career, and he never became reconciled to the idea of dele-

[50] *Ibid.*, 391-392. See also J. S. Bassett, *Slavery in the State of North Carolina*, 8.

[51] H. R. Helper, *Impending Crisis*, 218.

[52] *Arator*, I (November, 1855), 249.

[53] J. S. Bassett, *Slavery in the State of North Carolina*, 7-9, 42; J. G. deR. Hamilton, *Party Politics in North Carolina, 1835-1860*, 206-207; William Edward Dodd, *The Cotton Kingdom*, 68-69; L. C. Gray, *History of Agriculture*, I, 508-528.

gating his responsibilities or authority to a hired man.[54] With most of the large planters the problem was not so easy of solution. They needed assistance in supervising and executing their plans and employed overseers for that purpose.[55]

"Franklin" wrote in 1840: "Of all the difficulties of a planter's life, the greatest beyond comparison, is to procure and retain a good overseer."[56] When one considers the duties and responsibilities of "this useful and necessary class of men," as this correspondent called them, there can be little doubt concerning the accuracy of the above statement. For the responsibility they bore and the work expected of them, they were probably the most underpaid workers in all our economic history.[57] Thomas Ruffin, Junior, wrote his father about his efforts to employ a new overseer, of good reputation, but whose terms were $225 yearly and 500 pounds of meat: "I told him you would hardly give so much, but could not tell until I heard from you."[58] Ruffin was a good business man and is not to be considered penurious by any means; yet, this employee was to execute the owner's plans for the operation of a plantation having over one hundred slaves and from which the owner must of necessity be absent most of the time! The salary mentioned, however, was in keeping with the prevailing rate of that time. Ebenezer Pettigrew employed one man as his overseer for one dollar per day and at the same time hired a slave belonging to this man for ninety cents per day![59] "Franklin" proposed that the planters come to an understanding concerning contracts with overseers. He recommended a salary scale be adopted of from $300 to $600 per annum.[60] By the latter part of the period most planters had abandoned the share-crop system in favor of a money wage in paying their overseers.[61] Because of the greater skill

[54] B. H. Wall, "Ebenezer Pettigrew," 183, 192.

[55] For a discussion of the labor organization of the plantation, see Walter L. Fleming, "The Slave Labor System in the Ante-Bellum South," *The South in the Building of the Nation,* V, 107-108. The letters of Moore Rawls, overseer for Lewis Thompson of Bertie County on his Louisiana plantation, give a very good insight into the character of these agents and the weight of responsibility they bore—especially in the absence of the plantation owners. Lewis Thompson Papers, Southern Historical Collection, Library of the University of North Carolina.

[56] *Carolina Planter,* I (February, 1840), 41. For an account of one planter's actions in this connection, see B. H. Wall, "Ebenezer Pettigrew," 180-218.

[57] For a discussion of the qualifications, responsibilities, salary, education, etc., of overseers, see J. S. Bassett (ed.), *The Southern Plantation Overseer,* 6-9. For a generous account of a particular overseer, see James Calvin Bonner, "The Plantation Overseer and Southern Nationalism As Revealed in the Career of Garland D. Harmon," *Agricultural History,* XIX (January, 1945), 1-11.

[58] J. G. deR. Hamilton (ed.), *Papers of Thomas Ruffin,* II, 561.

[59] B. H. Wall, "Ebenezer Pettigrew," 186.

[60] *Carolina Planter,* I (February, 1840), 41.

[61] John Taylor in his *Arator* severely condemned the practice of giving the overseer a portion of the crop in return for his services; he maintained that the practice had contributed a great deal toward the wastage of the soils of the South by placing the land under the burden of an "annual rack rent." Quoted in *Farmer's Register,* VIII (December, 1840), 719-720.

required and the more unfavorable working circumstances encountered, the pay of overseers on rice plantations was probably higher than elsewhere.[62]

Practically every farm paper at one time or another carried a list of rules or duties for overseers.[63] These, of course, had about as many variations as there were employers, but, in general, they set forth the purpose of the owner and bound the overseer to "mak Mr. Allston's interest my own," as Duncan Barrantine of Anson County put it.[64] Usually the owner did the planning and the overseer executed those plans.

That some planters were more fortunate in their selection of overseers than others goes without saying. Thomas Ruffin's son wrote him as follows:

> Your overseer manifests the same . . . zeal in your service that he did when I wrote you before. The best Painter could not depict, nor the most skillful statuary carve, a face more indicative of grief and despair than is his, when one engages in conversation with him upon the weather and the state of the farm.[65]

Ebenezer Pettigrew could not have hoped for a greater measure of loyalty than that freely given by one of his overseers when he thrashed his own friend for speaking disrespectfully of Pettigrew.[66]

James H. Ruffin, of Caswell County, a lawyer who left the management of his plantation to his overseer, Dobbs, wrote his brother Thomas Ruffin as follows:

> . . . this man Dobbs will *ruin* and *beggar* me if he is not more closely watched than he has heretofore been. He is a man in my opinion totally devoid of principle and without the smallest regard to my interest when his own is not also consulted. He has under his management and control every thing I am worth to use or to abuse it according to his own will and pleasure.[67]

Most of the overseers were either landless whites or poor farmers, and, to a considerable extent, they were illiterate, superstitious, conservative, and unacquainted with and opposed to the ideas being proposed by the "book farmers." Of necessity, however, they usually possessed

[62] J. Harold Easterby (ed.), *The South Carolina Rice Plantation: As Revealed in the Papers of Robert F. W. Allston*, 23-27. Cited hereinafter as *The South Carolina Rice Plantation.*

[63] *Carolina Planter,* I (July, 1840), 203; *DeBow's Review,* XVIII (March, 1855), 339-345; *Farmer's Register,* VIII (April, 1840), 230-231; J. S. Bassett (ed.), *The Southern Plantation Overseer,* 11-22; U. B. Phillips, *American Negro Slavery,* 261-290. One contributor recommended that a prize be given for the best essay on the duties of overseers "in the form of an address, to persons engaged in that employment." *American Farmer,* VIII (May, 1826), 60.

[64] J. H. Easterby (ed.), *The South Carolina Rice Plantation,* 276.

[65] J. G. deR. Hamilton (ed.), *Papers of Thomas Ruffin,* II, 25.

[66] B. H. Wall, "Ebenezer Pettigrew," 199-200.

[67] J. G. deR. Hamilton (ed.), *Papers of Thomas Ruffin,* I, 384-385.

a good measure of practical intelligence. Some few became overseers in preparation for later careers as planters. Both Thomas Ruffin and Ebenezer Pettigrew employed their sons with that purpose in view. Generally the planter looked upon the overseer as a person without social standing—a hired servant withal—and he was dealt with as such. To the slave he was the symbol of all that was despicable in the slave institution. He summoned them to work in the morning, he passed judgment on their work during the day, he was the strong arm of the master in the enforcement of plantation discipline, and ever stood athwart the path "of any slave who had liberal ideas of the comforts of bondage." As Bassett phrased it: "The overseer was not loved: as a rule he was not lovable."[68]

To assist the owner or overseer in the management of the labor, there was selected from among the slaves one or more "drivers," whose duties were to both lead and drive the other workers.[69] They assisted in turning the slaves out in the mornings, leading them to work for the day, laying out and assigning tasks for individuals or groups, caring for livestock and equipment, checking on completion and quality of work performed; and, in all, served as a sort of sub-overseer. Indicative of the special concern the planter felt for this servant is the following rule which was proposed by a planter for the overseer's guidance: "The overseer must not punish the driver except in some extraordinary emergency that will not allow of delay, until the employer is consulted. Of this rule, the driver is, however, to be kept in entire ignorance."[70]

Three systems were used in working the slaves. The most popular and perhaps the best was that whereby each worker was assigned a task for the day or week, which, upon satisfactory completion, gave the slave the rest of the day or week off. The second was the gang labor system wherein the driver "speeded up" the tempo of activity by setting the pace himself and insisting that the gang keep up. Lastly, and no doubt the poorest, was the system whereby the slave was sent to work without any supervision or control except the fear of punishment.[71]

In the latter days of slavery, when more thought was being given to the efficiency and profitableness of the institution, many correspondents contributed estimates to the farm journals of what they considered

[68] J. S. Bassett (ed.), *Southern Plantation Overseer,* 3-8. A good appraisal of overseers generally is contained in *ibid., 22, 274-275.* The overseer's task was undoubtedly difficult. As Moore Rawls, overseer for Lewis Thompson of Bertie County, put it: ". . . men of the right stamp to manage negroes are like Angels visits few & far between. . . ." Moore Rawls to Lewis Thompson, undated, Lewis Thompson Papers.

[69] For an account of the origin of the term and the duties of the drivers, see Duncan Clinch Heyward, *Seed from Madagascar,* 157-158.

[70] *Farmer's Register,* VIII (April, 1840), 231.

[71] J. A. C. Chandler and others, (eds.), *The South in the Building of the Nation,* V, 107-108.

suitable "tasks for able bodied labourers," covering the manifold activities into which slave labor was directed.[72]

North Carolina had a considerable number of free Negroes in the period, many of whom were employed as laborers on farms and plantations. Generally, they were unwanted and more and more restrictions were imposed on their freedom. Calvin Jones wrote in 1828: "As for the free Negroes, they are a degraded race, below the slaves in all those qualities which procure happiness for themselves or produce usefulness to the community."[73] Many slaveowners felt that the presence of free Negroes in the community exercised an unfavorable influence on the slaves. John Taylor of Caroline was one of those. Taylor would deny permission to all free Negroes to enter a slave community and require those slaves receiving freedom by manumission to leave immediately.[74]

In 1826, North Carolina passed a law prohibiting entry into the state of any free Negro or mulatto with penalty of $500 for violation, which, if not paid, also required that the offender be sold into slavery for a period of not to exceed ten years.[75] The city of Wilmington passed an ordinance in 1857 which prescribed a $50 fine on any free Negro who came ashore from any ship that might be in the harbor without the expressed permission of the mayor.[76] In the same year, one planter described the free Negro as a nuisance and proposed that a petition be made to the legislature that all free Negroes be required to leave the state within twelve months "or hunt their owners; and if they refuse, put them up to the highest bidder, and let the money be added to the common school fund."[77]

Notwithstanding the restrictions under which they lived, free Negroes in the state increased from about 5,000 in 1790 to about 30,000 in 1860—a ratio of increase greater than that shown by the white people. Most of these people lived in and around the small towns, however, many of them found employment as agricultural laborers.[78]

The landless, slaveless white people, usually referred to as "poor whites," who were to be found in nearly every community in the state, provided another source of labor for agriculture. The ease of acquiring land and the fact that they were in competition with slave labor not only kept the number in this class small but their wages low. In 1850, the

[72] *Southern Agriculturist*, VI (November, 1833), 587-591; Francis S. Holmes, *The Southern Farmer and Market Gardener*, 230-232.

[73] Quoted in *American Farmer*, X (November, 1828), 273.

[74] See reprint of Taylor's *Arator* in *Farmer's Register*, VIII (December, 1840), *passim*.

[75] *Laws of N. C.* (1826), 13-16.

[76] *North Carolina Planter*, I (October, 1858), 304.

[77] *North Carolina Planter*, I (October, 1858), 304. For a comprehensive study of the restrictions imposed upon the free Negro in ante-bellum North Carolina, see John Hope Franklin, *The Free Negro in North Carolina, 1790-1860*, 14-57, 192-221.

[78] Norfleet Papers, *passim*. For excellent accounts of the free Negro in ante-bellum North Carolina, see G. G. Johnson, *Ante-Bellum North Carolina*, 582-612; J. H. Franklin, *The Free Negro in North Carolina, 1790-1860*, *passim*.

average monthly wage of such a worker, with board furnished, was only $7.21.[79] By 1860, though there had been a tremendous increase in the price of slaves and agricultural conditions were prosperous, this figure had risen to only $10.37.[80]

Shiftless, unreliable, and antagonistic to the Negro, this worker most often performed odd-jobs service for those able or willing to employ him. Ebenezer Pettigrew used them to belt trees, clear new ground, hunt stray livestock, hunt runaway slaves, cut and roll logs, take slaves to and from jail, run errands, and fire brickkilns.[81] Stephen Norfleet used them in ditching.[82] Thomas P. and Henry K. Burgwyn, brothers and leading planters, came into possession of estates in the Roanoke River valley in 1840. Both had travelled extensively and planned to manumit their slaves and use hired labor in their operations. After losing considerable money hiring Irish laborers they gave up their ideas about manumission and prospered throughout the remainder of the period.[83]

Undoubtedly, slavery as a source of agricultural labor was growing in popularity in North Carolina in 1860. As the slave grew in efficiency as a laborer, free labor became less and less able to compete.

[79] J. D. B. DeBow, *Statistical View of the United States . . . Being a Compendium of the Seventh Census. . . .*, 164.

[80] *U. S. Census, 1860* (Statistics), 512. In the whole field of wage payments, North Carolina's average per worker ranked lowest in the Union in 1860.

[81] B. H. Wall, "Ebenezer Pettigrew," 174.

[82] Norfleet Papers.

[83] H. A. Kellar (ed.), *Solon Robinson*, II, 224 n. For additional comment on free white labor in North Carolina agriculture, see B. H. Wall, "Ebenezer Pettigrew," 151-179; F. L. Olmsted, *A Journey in the Seaboard Slave States*, 330; H. R. Helper, *Impending Crisis*, 380-382; G. G. Johnson, *Ante-Bellum North Carolina*, 68-73; R. H. Taylor, *Slaveholding in North Carolina*, 80; L. C. Gray, *History of Agriculture*, I, 500-501.

CHAPTER IV

# AGRICULTURAL IMPLEMENTS IN NORTH CAROLINA, 1783-1860

The development of agriculture in colonial North Carolina, as in the other English-speaking colonies, was characterized by widespread experimentation with crops, with modifications of European farming practices, and with the development of the slave-plantation economy.[1] In most respects the State's farming potential had been determined by 1783 and the general pattern of its present-day agriculture fixed. Even in that day the farmers in the eastern third of the State tended to operate on a large scale, producing crops for a commercial market, whereas those in the middle and western parts of the State more often were engaged in small-scale, self-sufficing farming.

In one significant respect, however, the colonial period was almost lacking in accomplishments. Scarcely any new tool or farming implement was added to those which the earliest settlers had brought from Europe. In fact, with the exception of a "machine," similar to the horse-hoe, which was observed in use on a colonial North Carolina tobacco farm,[2] the farmer's implements were strikingly similar to those referred to in the treatises of Cato and Varro concerning the agriculture of ancient Rome.[3] Despite the contrast in conditions under which farming was carried on in Europe and America, the colonial farmer continued to use implements with which he had been long familiar.[4]

Every inventory of colonial implements included some of the following: augers, axes, chisels, hammers, wedges, saws, harrows, hoes, plows, spades, shovels, picks, mattocks, sickles, scythes, reaping hooks, forks, flails, rakes, wheelbarrows, carts, and wagons.[5] There is considerable evidence that most farmers lacked some of these tools and that farming in colonial America was carried on primarily with hand labor.[6]

[1] This chapter was presented at the session of the Agricultural History Society with the American Historical Association at Boston on December 30, 1949, and was published in the July, 1951 issue of *Agricultural History*. Republication here of the text of this article has the approval of the Editor of *Agricultural History*.

[2] Harry J. Carman (ed.), *American Husbandry*, 249.

[3] [Fairfax Harrison, ed. and tr.], *Roman Farm Management: The Treatises of Cato and Varro Done into English. . .*, 118-120.

[4] Jethro Tull's horse-drawn seed drill has been cited as the only new implement invented in either England or America during the colonial period. Lyman Carrier, *The Beginnings of Agriculture in America*, 263.

[5] L. C. Gray, History of Agriculture, I, 194; and William H. Gehrke, "The Ante-Bellum Agriculture of the Germans in North Carolina," *Agricultural History*, IX (July, 1935), 150-151.

[6] E. W. Andrews and C. M. Andrews (eds.), *Journal of a Lady of Quality*,

Skilled mechanics to reproduce implements and keep them in a good state of repair were scarce in the colony.[7] The average farmer, therefore, lacking easy access to British sources of supply, was compelled to rely upon his own craftsmanship to fulfill his requirements of implements. Consequently, his implements were usually inferior in quality and in number to those in use on European farms. This fact was remarked upon by visitors in colonial North Carolina.[8]

Needless to say, farming in the colony was characterized by inefficiency and carelessness. Also, that the scarcity of good implements contributed to such a condition. Fields which were only scratched over, which were not provided with adequate drainage, and in which hardly any use was made of crop rotation or cover crops soon lost their productivity and top soil. Land was the abundant factor in the production process and was given very little intelligent attention. By 1783, the practice of abandoning worn-out fields to grow up in bramble and briers had already marred the landscape. Perhaps North Carolina farmers were more negligent in these respects than others; for, by the end of the colonial period, North Carolina had acquired a reputation as the most backward, short-sighted, unintelligent, and unprogressive colony among the continental English group in matters pertaining to agriculture.[9]

Unfortunately, independence of England brought little prospect of improvement in the status of North Carolina farmers or in the quality of their work. In fact, importation of farming implements from abroad became more difficult, and local sources of supply were slow to develop. Both farmers and planters were generally compelled either to make their own implements or to rely upon the small number of blacksmiths who were to be found in the State.

The blacksmith became a particularly important member of the community and rendered invaluable service to the farming interests of the State by reproducing and repairing implements of all sorts. Some few acquired reputations for special skills, and their services were widely sought after.[10] Most blacksmiths, however, engaged in a general sort of practice. John Morrison of Raleigh, for example, solicited business in repairing coaches, gigs, stages, and wagons and in "making and repairing patent ploughs, and all other kinds of farming tools, shoeing horses, &c."[11] Undoubtedly the work of such artisans varied widely in quality, and much of it was crude in appearance and inefficient in operation.

163; L. C. Gray, *History of Agriculture,* I, 170, 365; W. Neil Franklin, "Agriculture in Colonial North Carolina," *N.C.H.R.,* III (October, 1926), 559-560.

[7] See the remarks of Bishop Spangenburg, a noted visitor in the colony in 1752, as recorded in *C.R.N.C.,* V, 1-14.

[8] For example, see the remarks in E. W. Andrews and C. M. Andrews (eds.), *Journal of a Lady of Quality,* 163.

[9] H. J. Carman (ed.), *American Husbandry,* 248-259.

[10] J. G. deR. Hamilton (ed.), *Papers of Thomas Ruffin,* I, 195. See also the advertisement of Robert Temple in the Raleigh *Minerva,* September 17, 1807.

[11] Raleigh *Minerva,* July 30, 1820.

Charges for blacksmith hire or services were very reasonable, For example, a farmer of Orange County, had twenty-four charges made against him for blacksmithing during the year ending in July 1833, for such services as "pointing a shovel," "refitting a spade handle," "pointing a plow share," "mending a clevis," and "ironing single-tree," for which the total charge was only $4.58½.[12] Also, frequent notations in ante-bellum plantation expense accounts of the item, "Blacksmith Hire," may be seen for which the wage varied from $1 to $1.50 per day.[13]

The planter was somewhat less dependent than the farmer upon his own exertions or those of a local blacksmith for his supply of implements. Having more funds usually and better trade connections with the more industrialized North, he was the first to learn of new implements and to import them into the State. Henry King Burgwyn, one of the leading planters in the State, for example, while on a visit to Boston in 1843 bought six large plows and six small ones which he shipped to his plantation in Northampton County.[14] In such manner, the best implements of that day, such as the Dagon, Cary, and Freeborn plows, and the Hussey and McCormick reapers were introduced into the State. Once available, such implements became the objects of much tinkering and modification by those who observed them in operation.

Not even the planters, however, had an adequacy of tools and equipment for their farm work. This fact was strongly emphasized by the advocates of agricultural reform who began to make their voices heard in both State and Nation after about 1800. "I have known considerable planters," wrote Charles Fisher of Rowan County, "instead of iron traces, make use of grapevines and hickory withes. What false economy! What miserable management!" Fisher, like most of the advocates of general agricultural reform, drew attention to the need for better farming implements. "Examine the tools of the greater number of our planters," he wrote, "and we need search for no other reason why our fields are only half cultivated."[15]

Agricultural societies, editors of farm journals and papers dealing with agricultural subjects, and progressive farmers all added to the agitation for and interest in better farming implements. Stimulated by their encouragement and the great need for improvement, inventive minds were set to the problems with the result that many farming implements or machines were improved or invented. The trend toward improvement in this respect may be seen in all directions. Between 1783 and 1860, more new implements were introduced into agriculture

[12] Burwell Benson Papers, Southern Historical Collection, Library of the University of North Carolina.

[13] For an example of such entries, see the Norfleet Diary, *passim,* Norfleet Papers.

[14] Henry King Burgwyn Papers, Southern Historical Collection, Library of the University of North Carolina, cited hereinafter as Burgwyn Papers.

[15] *American Farmer,* III (August, 1821), 169-170.

than had been done in the previous two thousand years. Also, many old implements were modified in a revolutionary manner. Never before had so much thought been given to the farmer and his problems.

North Carolinians were alert to these developments and had a significant part in bringing them about.[16] Although too numerous to note in detail, some of these inventions deserve special mention. In 1803, G. F. Saltonstall became the first North Carolinian to secure a patent for a new agricultural device. His was for a saw gin. In 1817, a traveler through the Piedmont counties of Burke and Lincoln observed several corn pounding machines in operation. In function, this machine was identical with the present-day gasoline or electrically driven hammer mill. Designed to crush both cob and grain for purposes of feeding cattle and hogs, it was ingeniously constructed for operation by water power. It was not patented, and the inventor is unknown.[17]

George W. Jeffreys, one of North Carolina's most prominent advocates of agricultural reform, published a description in 1819 of a drill plow, a horse rake, and a seed coverer which he had designed, built, and was using in his planting in Person County. These implements were simple in design and could be easily reproduced by following the instructions which he furnished.[18]

George W. B. Burgwin, of near Wilmington, described a horse-drawn seed coverer which he had made and was using in 1821 to cover his cotton as being superior to the hoe for that purpose and besides saved the work of six or eight hands.[19]

Ebenezer Pettigrew designed and made what he called a "mud machine" in the 1820s which he used in extensive dredging and land reclamation operations on his plantation in Tyrrell County.[20] He also invented a power-driven corn sheller, which he did not patent, and which he refused to commercialize upon. Pettigrew had a tremendous advantage over the average planter in that he had harnessed the outflow of water from Lake Phelps, on which his plantation was located, and was using water power in processing his harvests and in making and repairing his farm implements.[21]

One of the earliest diagrams of a farming implement carried in an American agricultural journal was that of the "Carolina" plow, "newly invented or improved," which appeared in the *American Farmer* in 1822. This plow was described by a contributor as being far superior to other

[16] The original patent law of 1790 gave inventors who patented their ideas exclusive rights to their discoveries. For a chronology of major American inventions during 1790-1860, see Roger Burlingame, *March of Iron Men*, 468-476.

[17] *American Farmer*, II (May, 1820), 47-48.

[18] Denison Olmsted, *Report on the Geology of North Carolina*, Part I, 83-84.

[19] *American Farmer*, III (August, 1821), 156.

[20] J. G. deR. Hamilton (ed.), *The Papers of Thomas Ruffin*, I, 434-435.

[21] For a good description of this plantation, see Edmund Ruffin in *Farmer's Register*, VII (November, December, 1839), 698-703, 724-733.

popular models of the day. It was light, strong, could be built and repaired by a common smith, and only cost $5. Sheet iron was used in the construction of the moldboard.[22] Also in 1822, Joseph Delaplane, an overseer by profession, reflected the increase of interest in deep plowing by inventing what he called a "substratum" plow. This implement was described as being sturdy, inexpensive, and simple in design.[23]

North Carolinians shared in the widespread efforts to produce satisfactory planters for cotton, corn, wheat, and other grains. Perhaps the best results were obtained by two carriage makers of Yanceyville who patented a "machine" for planting cotton in 1841. This implement attracted considerable attention and favorable comment. "The merits of the machinery," wrote the editor of one paper, "consists in its opening the furrow, sowing the seed, and covering it, which we . . . believe it does . . . with as much ease, as simply ploughing a furrow."[24] This implement was designed in such a manner that it could also be easily converted to plant corn—either in drills or in checks.

Gray Utley, of Chapel Hill, patented a straw cutter in 1845 which received much praise from the editor of one agricultural paper and other persons who examined it. This machine could be used to cut stalks as well as straw, cost from $12 to $15, and was recommended as a "valuable invention."[25]

Between 1850 and 1860, the most prosperous decade for ante-bellum North Carolina farmers, the number of inventions or improvements in farming implements was rapidly increased. In 1850, Joel K. Holland of Washington patented a manure cart which could be used in distributing manures either broadcast or in drills, or as an all-purpose cart. Whitman Price of Wayne County invented a cotton thinner and cultivator which, according to an enthusiastic newspaper editor, "with one horse and hand, performs the labor of two horses, and six to eight hands."[26] This machine was designed to plow the young cotton on both sides and thin it to the desired distance all in one operation. Price planned to sell county and State rights to produce this implement, but there is no evidence that he had any success in that direction or that the machine received more than local interest. Price also patented a potato planting plow, a turn plow, and a cotton planter.[27]

Implements which could be made, by adjustments, to serve a variety of purposes attracted a great deal of attention. W. E. Wyche of Granville County patented what he called a "mole plow," which could be used in either shallow or deep plowing, siding, and listing.[28] Alexander

[22] *American Farmer,* IV (April, 1822), 31.
[23] Hillsboro *Recorder,* June 12, 1822, quoting the *American Farmer.*
[24] Tarboro *Press,* February 6, 1841, quoting the Salisbury *Carolina Watchman.*
[25] *North Carolina Farmer,* I (November, 1845), 143.
[26] Tarboro *Southerner,* June 24, 1854.
[27] *Report of the Commissioner of Patents* (1854), (1860).
[28] *Arator,* II (March, 1857), 723.

Dixon and William B. Williams also patented all-purpose plows. Of the Williams' implement, the editor of the *North Carolina Planter* wrote: "We are decidedly of the opinion that this is the best and cheapest plow in the state. It is eight plows in one. . . ."[29]

A corn-husking machine patented by F. M. Walker, of Greensboro, was described as "strong and simple and not liable to get out of repair." With this machine, it was said, one man could husk from 35 to 40 barrels of corn in a day.[30]

Many improvements in implements and inventions of new ones were produced by North Carolinians during the ante-bellum period which were never formally recorded in the United States Patent Office, but which were used in farming. From 1790, when the first patent law was passed, to 1850, 33 patents were issued to North Carolinians for inventions of agricultural implements and 20 for implements or devices in the category: "Grinding mills and mill gearings, containing grain mills, mechanical movements and horsepowers."[31]

The progress of the agricultural reform movement, which was sweeping over the State in the 1850s, was reflected by an increase in the number of patents issued. In all, North Carolinians secured patents in that decade for the following farm implements: 1 corn husker, 1 corn harvester, 3 cottonseed planters, 1 cotton-thinning plow, 1 cottonseed cleaning machine, 1 stalk cutter, 4 straw cutters, 7 plows, 10 cultivators, 2 seed planters, 1 hillside plow, 3 sowing machines, 1 harvesting machine, 1 thresher, 1 potato-planting plow, 1 wagon brake, 4 manure carts, 1 feeding apparatus for a grain thresher, 1 self-dumping truck, 1 marl lifting machine, 1 excavator, 1 hand truck, 1 water wheel, 1 pump, 1 carriage spring, 1 hoof paring knife, and 1 machine for stuffing horse collars. In anticipation, perhaps, that such labor-saving devices would provide more leisure time for the farmers one ingenious North Carolinian secured a patent for a "self-waiting" table and another for a "ventilating" rocking chair.[32]

Surprisingly, there is no evidence of attempts to produce any of these implements on a large scale; or that any inventor profited greatly from his invention.

Agricultural reformers, societies, and editors of farm journals drew attention to the improvements being made in farming implements and strongly endorsed them for general use. Farm papers began to carry diagrams of new plows, carts, reapers, cultivators, horsepower machines, marl diggers, etc., accompanied by the enthusiastic endorsements of the

[29] *North Carolina Planter*, III (November, 1860), 358.
[30] *Ibid.*, I (May, 1858), 160.
[31] *Report of the Commissioner of Patents* (1849), 126.
[32] *Ibid.* (1850-1860), *passim*.

editors.[33] Almost without exception, the economies which would result from investments in improved implements were emphasized. The editors also sought to drive this lesson home by frequent quotation of what were called the "Fixed Facts of Agriculture." The following are typical quotations: "Full complements of tools, and implements of husbandry, are intimately connected with the success of the husbandman"; and "Capital is not only necessary to agricultural success, but can be as profitably used in farming, as in any other occupation."[34]

In their zeal for the agricultural cause, editors often rendered a disservice by their too hasty or overly enthusiastic endorsements of new and untested implements. Such a practice was called "puffing." The editor of the Tarboro *Southerner,* for example, said of a newly invented cultivator, being announced in his paper but which he had not seen: "It must be one of the most valuable labor-saving implements ever invented."[35] Edmund Ruffin, in his *Farmer's Register,* was far more restrained in this respect than most and exercised a helpful toning-down influence on his more gullible brethren.[36]

Henry King Burgwyn, one of North Carolina's most successful planters and heaviest purchasers of farm implements and machinery, roundly condemned the editors of farm papers for their extravagant praise "so often bestowed upon a worthless implement." Such extravagant praise, he said, would evoke from the farmer suspicion and an overly cautious attitude especially when he knew that the implement being praised was worthless. Burgwyn submitted for publication a long list of implements which he had used, calling each by name, indicating where he had gotten it, what it had cost, and giving his opinion of each. For example, he said of a horsepower wheat thresher, patented by Taplin and highly recommended by the *American Agriculturist,* which he had bought from E. Whitman and Company of Baltimore: " . . . it was an unmitigated failure, often losing three days of time. . . .", and 500 bushels of wheat were lost in trying to make it work. Burgwyn urged other farmers to *"do their duty,* and come out under their own signatures" with appraisals of the implements they were using. Although he was perhaps as free from sectional prejudice as any planter in North Carolina he accused Northern manufacturers of being careless as to quality in those implements produced for the Southern market,[37]

Toward the end of the ante-bellum period, the enthusiasm with which

[33] A. L. Demaree, *The American Agricultural Press 1819-1860,* 46-47. For a good illustration of the type of support given a farm implement concern, see the editor's remarks concerning the William H. Oliver and Co. of Newbern in the *North Carolina Planter,* I (April, 1858), 117.

[34] *Farmer's Journal,* I (August, 1852), 132.

[35] Tarboro *Southerner,* June 24, 1854.

[36] For an example of Edmund Ruffin's insistence that inventors and other promoters of new implements tell only the truth about their discoveries, see J. G. deR. Hamilton (ed.), *Papers of Thomas Ruffin,* III, 52-53.

[37] Tarboro *Southerner,* June 25, 1853, quoting the *American Farmer.*

new implements were announced was less wholehearted, particularly if the implement was of Northern or "yankee" manufacture.[38]

Agricultural reformers joined in attacking the farmers' prejudices against innovations in agriculture and in urging upon them the importance of having the best of implements. John Taylor, of Caroline County, Virginia, who through his *Arator* exercised a strong influence on agricultural reform in North Carolina, referred to agricultual mechanics as one of the neglected phases of that interest.[39] General William R. Davie, who also served his State as a critic of her farming practices, said that the evils in agriculture, such as "indiscriminate and shallow ploughing" could not be remedied until improvements had been made in farming implements.[40] Charles Fisher, in an address to the Rowan County Agricultural Society in 1821, spoke of "the deficiency in our farming implements." "It is as impracticable for the farmer," he said, "as it is for the mechanic, to do good work with bad tools."[41] Paul C. Cameron voiced a similar sentiment before the Orange County Agricultural Society and urged: "Make yourselves the owners of the best ploughs, teams, and carts. . . ."[42]

The rather general neglect in the care and maintenance of farming implements also came in for a sound condemnation. Charles Fisher told his Rowan County audience: "In this particular, the greater part of our farmers are culpably negligent. Go to one of their farms, and you may see a plough lying in this fence corner—in that a hoe; here one thing, and there another."[43] To combat such carelessness, Fisher and other reformers emphasized that the farmer should have a place for the care and safekeeping of all tools when not in use. "Have a place for every tool," advised another writer, "and never leave one out of its place; or to go further, a place for everything and everything in its place."[44] Such ideas were stressed in all the farm papers and by the end of the ante-bellum period, tool houses and sheds had become common features of the better farmsteadings.

Perhaps the most effective promotion given new farming implements and machinery was accomplished at the fairs which were sponsored in the State by various agricultural societies. Most of these societies had standing offers of premiums for the improvement of old implements or the invention of new ones. Some like the Mecklenburg County Agricultural Society, prescribed that, in order to win the premium, the implement's merit had to be demonstrated at the county fair.[45] The first

[38] *Farmer's Journal,* I (July, 1852), 116-117.
[39] *Farmer's Register,* VIII (December, 1840), 755-757, quoting the *Arator.*
[40] *American Farmer,* I (October, 1819), 218.
[41] *Ibid.,* III (August, 1821), 170.
[42] *Arator,* I (June, 1855), 73.
[43] *American Farmer,* III (August, 1821), 170.
[44] *North Carolina Planter,* I (January, 1858), 32.
[45] Raleigh *Standard,* June 4, 1845.

such exhibit held in the State was by the Rowan County society in 1821. Among the implements exhibited at that fair were a homemade two-horse plow, described as of excellent workmanship, an Enfield plow, and a cotton cultivator. The latter attracted much attention and approval from the visitors. The premium committee commented that "much labor might be saved by the use of this implement in the cultivation of cotton. . . ."[46]

The State fair, sponsored by the North Carolina Agricultural Society and held annually at Raleigh, beginning in 1853, offered sizable premiums in each of the following classes of implements: plows, harrows, threshing machines, corn shellers, reapers, mowers, hand implements, carriages and wagons, farming and domestic tools, and saddlery. In 1855, the general assembly began making annual appropriations to the premium fund. All railroads, except one, leading to Raleigh also offered free transportation for all exhibits to be shown at the fair.

These inducements led to an increase in participation in that feature of the fair. In 1858, items were exhibited in each of the ten premium classes—many of which were manufactured in the State. The judges were particularly attracted to a cotton cultivator "about to be patented" by a resident of Edgecombe County.[47] Of the implements exhibited at the fair that year, one editor wrote: ". . . we may say at none of the preceding fairs has near so many articles, of our own state manufacture, been exhibited. Everything needed on a Farm, it seems to us, can be procured in our own state of our own manufacturers."[48] At each succeeding fair the number and variety of implements exhibited exceeded the record of the previous year.

To meet the increasing demand for more and better farm implements, several concerns were established in the State during the ante-bellum period which specialized in the manufacture of farm implements of nearly every sort. The cotton gin early attracted the interest of enterprizing craftsmen. In 1807, Robert Temple, of Rogers Cross Roads, located about 14 miles northeast of Raleigh, was engaged in making cotton "machines," "constructed in the best manner, agreeable to Whitney's Patent," and selling them at a moderate price.[49] By 1810, there were 739 cotton gins in operation in the State.[50] It is probable that most of those were of local manufacture. Norfleet Tyer of Pitt County, Samuel D. Proctor of Wake County, and J. H. Thompson of Davidson County

[46] Salisbury *Western Carolinian*, October 23, 1821.

[47] *North Carolina Planter*, I (November, 1858), 352-353.

[48] *Ibid.*, 349. The United States Census of 1860, however, did not give North Carolina credit for having a single establishment producing shovels, spades, forks, hoes, scythes, and cotton gins! *United States Census of 1860* (Manufactures), ccxvi.

[49] Raleigh *Minerva*, September 17, 1807. For an interesting account of Whitney's fight to safeguard his patent rights to the cotton gin, see Jeannette Mirsky and Allan Nevins, *The World of Eli Whitney*, 111-127.

[50] Raleigh *Star*, February 21, 1811.

were other manufacturers of gins in the period.[51] Thompson also produced several kinds of plows, cultivators, and harrows and engaged in the general repair of implements.

Although the demand for a plow factory was voiced early, no factory was established in the State until late in the period.[52] Meantime most of the plows used were built by local blacksmiths or brought in from other States. In 1858, the Warrenton Foundry was specializing in the manufacture of plows and was turning out forty each week. "Every part of the work," wrote the editor of the *North Carolina Planter,* "is done on the spot . . . with all the advantages that steam-power, an excellent foundry, and skilful mechanics can produce."[53]

Most plants, equipped as well as the Warrenton Foundry, engaged in less specialized production. A machine shop was opened at Charlotte in 1858, for example, in which it was said anything could be made "from a steam engine to a horse shoe."[54] In 1856, the Clarendon Iron Works in Wilmington, which had been producing farming implements, was purchased by A. H. Vanbokkelen who began soliciting orders for steam engines, "any power or style," sawmills, gristmills, flour mills, water wheels, rice-field pumps and engines, corn and cob crushers, rice threshers, cotton gins, gearings, boilers, and blacksmith work of all kinds.[55] At the State fair in 1860, a similar concern, Frercks and Raeders, had on exhibit cultivators, plows, corn shellers, seed sowers, horsepower machines, threshers, separating and cleansing machines, and tobacco presses and fixtures which had been made in their foundry and machine shop at Salisbury.[56] Businesses engaged in the manufacture of carriages, harness, and weeding hoes were operated at Fayetteville, Tarboro, and Greenboro, respectively, in the decade of the fifties.[57]

Out-of-state producers of farming implements recognized early the advantage of either selling patent rights which would permit the manufacture of their implements in North Carolina, or of designating agents to promote the distribution of their products in the State. In either case, the interest of the reading public was cultivated by advertisements which began to occupy more and more space in newspapers and agricultural journals. Gavin Alves, for example, purchased in 1807 the patent rights to a corn shelling machine for distribution in Orange and Chatham counties. Alves, in his advertising, addressed himself to those planters "to whom the saving of time and labor is of importance," and presumed

[51] Tarboro *Press,* June 2, 1838; Salisbury *Carolina Watchman,* December 26, 1850.

[52] *American Farmer,* III (May, 1821), 53.

[53] *North Carolina Planter,* I (December, 1858), 387.

[54] *Ibid.,* I (July, 1858), 232, quoting the Charlotte *Democrat.*

[55] Tarboro *Southerner,* January 5, 1856.

[56] *North Carolina Planter,* III (November, 1860), 358.

[57] Tarboro *Press,* April 21, 1851, quoting the Fayetteville *Observer;* Tarboro *Southerner,* July 15, 1855; *North Carolina Planter,* I (March, 1858), 95.

that they would buy the machine, "as few will wish to employ a whole day in what may be accomplished in an hour."[58]

The sale of reapers, threshers, and horsepower machines, which were invented and gradually improved after about 1830, was promoted in the State by agents. Harman's threshing machine was one of the first to attract general interest. By 1833, three men had bought the patent rights for that machine and were engaged in its sale in the eastern part of the State.[59] In 1844, William J. McElroy, of Davie County, announced that he was sole agent in the State for William Kirkpatrick's "Portable Horse Power and Threshing Machine." This machine was offered for $150 delivered at Fayetteville or $170 delivered at his residence.[60] John A. Boyden, of Salisbury, secured the agency in western North Carolina for Hussey's reaper and mowing machine, which was one of the most satisfactory implements of its type. Hussey visited Davidson County and demonstrated his machine on the farm of Doctor William R. Holt, one of his satisfied customers and later president of the North Carolina Agricultural Society. The machine cut the grain as fast as nine or ten hands could bind it.[61] The editor of a local paper witnessed the demonstration and confessed to his readers that ". . . though somewhat given to doubting new things ourselves, we cannot doubt, in this case, the evidence of our eyes."[62] Henry King Burgwyn, who together with his brother, Thomas Pollock Burgwyn, produced in Northampton County perhaps the largest small grain crops in the eastern United States in the 1850s, spoke of the Hussey reaper as "the greatest boon to the Farmer they have had for years, in appliances of agriculture."[63] Burgwyn had used McCormick's reaper but considered it "a miserable failure." He and his brother maintained a supply of fifteen reapers to harvest their crops.

Although manufacturers of farming implements continued to market their products through agents located in the State, implement supply houses were set up toward the end of the ante-bellum period in all the leading towns. Through advertising in local papers, participation in agricultural fairs, and by other means they brought their wares to the attention of the public. The William H. Oliver and Company, located in Newbern, for example, got out a catalog in 1858 in which engravings of their stock were carried. Among the items offered for sale were a large variety of plows, straw and cornstalk cutters, corn shellers, harrows, cultivators, rice mills, crushers, horse powers, cotton gins, reapers, mowers, threshers, smut cleaning machines, and portable sawmills.[64]

[58] Raleigh *Minerva,* June 4, 1807.
[59] Tarboro *Press,* June 2, 1838.
[60] Salisbury *Carolina Watchman,* June 1, 1844.
[61] *Ibid.,* May 31, 1855.
[62] *Ibid.,* June 21, 1855.
[63] Tarboro *Southerner,* June 25, 1853, quoting the *American Farmer.*
[64] *North Carolina Planter,* I (April, 1858), 117.

From the above, one might be inclined to conclude that ante-bellum North Carolina farmers were adequately supplied with the most up-to-date farming implements. Unfortunately, such was not the case. The supply of good sturdy implements from all sources was never adequate. Also, the forces of conservatism, prejudice, ignorance, superstition, and the lack of enlightened self-interest persisted in the population and continued to retard general agricultural reform. It was no mere happen-so that Frederick Law Olmsted, who traveled through the State in the 1850s, found a North Carolinian farming with what Olmsted called "a most absurd little plow"; or that Edmund Ruffin found some of the best farmers and largest wheat growers in the State harvesting crops of as much as 5,000 bushels with a sickle. Most North Carolina farmers were using just such implements.[65] "We go on in the habits and with the ploughs of the first settlers," wrote Charles Fisher, "regardless of what is passing around us in the way of improvement. . . . Nothing so retards the progress of improvement as prejudices, and the attachment to old habits."[66] "Alpha," of Edgecombe County, where perhaps the State's best farming was being done, lamented in 1853 that the farmers had no implements "but the crudest, and that is kept prostrate by importations of trashy productions of northern workshops."[67]

Notwithstanding the fact that many inferior and outmoded implements were continued in use in North Carolina agriculture, the better farmers of the State were alert to the improvements being made and were improving their farming by using the best of implements. This is particularly true of the last years of the ante-bellum period. Between 1850 and 1860, the rate of increase in the capital invested in farming implements and machinery in North Carolina was greater than for the Nation at large. North Carolinians were using more and better farming implements than ever before. Prejudice against innovations was gradually overcome and the foundation established for a widespread acceptance of the new devices.

[65] F. L. Olmsted, *Journey in the Seaboard Slave States,* 320-321; Edmund Ruffin, *Agricultural, Geological, and Descriptive Sketches of Lower North Carolina and the Similar Adjacent Lands.* 85.

[66] *American Farmer,* III (August, 1821), 170.

[67] *Farmer's Journal,* I (October, 1853), 211.

CHAPTER V

# AGRICULTURAL REFORM, 1820-1860

As previously noted, agriculture in North Carolina was in a crisis in 1820.[1] Because of a combination of circumstances and conditions a "state of embarrassment never before equalled in North Carolina" had come to pass. This condition was manifested by an increasing exodus from the state of people from all ranks of society, thousands of abandoned or wasted acres, depressed land values, low prices of farm products and a decreasing margin of profit in cotton and tobacco raising because of western and foreign competition, poorly developed marketing and transportation facilities, widespread poverty and ignorance, low morale, and an "unfavorable" balance of trade against the state.[2] To overcome the practices and conditions responsible for that state of affairs became the goal of both agricultural and political leaders. To note the programs of those leaders and to measure the degree of their success, as pertains to agriculture, is the purpose of this chapter.

The first impulse toward a renaissance of agriculture came from individuals.[3] What the agriculture of the state needed, according to Denison Olmsted, was not a new crop but that it become "an object of thought," so that more pains and intelligence would be used in the culture of the old crops.[4] Paul C. Cameron observed that all efforts to reform agriculture would be in vain unless farmers themselves developed a stronger will and a fixed determination to excel.[5] Individual reformers, therefore, undertook early to organize those agriculturists who were alert to the need and interested in reform.

The county and state agricultural societies formed in the state after 1810 played a significant part in educating the public to the need for reform, and, in their programs, indicating the direction that reform should take.[6] A similar movement was underway in the nation at the same time.

[1] See Chapter II, *supra*.

[2] Elisha Mitchell, *Report on the Geology of North Carolina. . . .*, Part III, 45-48.

[3] "M. N. E." in Hillsboro *Recorder*, August 21, 1822.

[4] *American Farmer*, VI (December, 1824), 318.

[5] *Arator*, I (June, 1853), 73. Cameron, whose principal plantation was in Orange County, but who also had a plantation in Mississippi, was an outstanding farm leader in the latter part of the period.

[6] Helen D. Wilkin, "Promotion of Agriculture in North Carolina, 1810-1860," 14-50, 80-141, unpublished masters thesis at the University of North Carolina (1941); E. L. Bogart, *Economic History of the American People* (2d. ed.), 307; J. D. B. DeBow, *Industrial Resources*, I, 72-77; J. A. C. Chandler, and others, (eds.), *The South in the Building of the Nation*, V, 552 ff.

Though more active than most, the Rowan County Agricultural Society, organized in 1821, may be considered typical of the county organizations.[7] Like most of these societies, its membership consisted of "a number of respectable planters of the county." Charles Fisher became the first president. A constitution was drawn in which provision was made for two meetings annually, at one of which the society-sponsored "Agricultural Exhibition and Cattle Show" was to be held.[8] The duty of a member as set forth in this document was

> diligently to turn his attention (as much as his situation will allow) to the study of agriculture, and on all occasions to impart to the society any improvements or discoveries he may make; also, to use every exertion in his power to procure correct models of the most approved farming implements in use in any part of the country.[9]

A fair, sponsored by this society in 1821, was described by a local newspaper as "being novel in this part of the country," and "of course was not so generally understood by the farmers as it will be in the future."[10] There were exhibited the following animals: five colts which had been foaled in the county; a large English Short Horn bull, for which "Mr. Jones" was awarded "An honorary premium . . . for his public spirit in bringing an animal of such superior quality into the county"; a young "half-breed" bull; and two Chinese boars. Of the implements exhibited, there was a home-made two-horse plow which was described as of "superior workmanship"; an "Enfield plow, No. 9"; and, a cotton cultivator which attracted much attention, and the observation that "much labor might be saved by the use of this implement. . . ."[11] Significant among the vegetables on display was the mangle wurtzle, then being highly recommended throughout the country for use as food for cattle.[12] A silver cup, valued at $10, was awarded for the "most approved practical Essay on the subject of manures, particularly vegetable and atmospheric." A similar cup was given for the best essay on the subject of raising livestock "in which the errors and defects of the present practices must be plainly pointed out and a better plan recommended." Announcement was made that a $25 award would be made at the fair in 1824 for the "neatest and best live fence or hedge" which had to be accompanied by "a particular account" of the methods used in its culture.[13]

[7] Salisbury *Western Carolinian,* July 10, 1821.
[8] *American Farmer,* III (August, 1821), 158-159.
[9] *Ibid.,* 163-164.
[10] Salisbury *Western Carolinian,* October 23, 1821.
[11] Salisbury *Western Carolinian,* October 23, 1821.
[12] *Ibid.*
[13] *American Farmer,* III (August, 1821), 158-159. Though a feature of practically every fair and agricultural society, there is little evidence of widespread competition for the awards offered for essays. Most of the winners were well educated planters or professional men. In this respect, the generality of planters

The North Carolina Agricultural Society was organized in 1818 and held annual meetings in Raleigh for a number of years thereafter. This body sought to improve agriculture by offering a large number of awards for essays on selected topics. Significant among those offered were $15 for the most productive meadow, "rendered so by art"; $15 for the best corn, wheat, rye, or cotton crop produced on a plot of two acres or more which had been exhausted and reclaimed; $10 for the best plow capable of deep plowing; $30 for the best barn built since 1818; and, $15 for the best account of the cattle distemper with an effectual method of preventing and curing it. The time was not thought to be right for holding a state fair and provision was made that in order to win the awards offered, the contestants "should furnish the best evidence the nature of the case will admit of their claim."[14]

In 1821, the state society decided to promote the establishment of similar organizations in each county in the state as auxiliaries of that body, and to receive delegates from the county societies at its annual meetings. It was also decided to continue offering premiums for essays on selected subjects, and to select and publish those considered best in the hope of stimulating agricultural advance. The titles proposed for these articles "The Best Method of Reclaiming Worn-Out Lands," and "The Production and Application of Manures Using Native Materials for Native Crops," indicate the direction it was thought agricultural reform should take.[15]

In 1822, the legislature made an appropriation of $10,000 to be divided among the counties in proportion to their federal numbers, for the purpose of developing agriculture and domestic manufacturing. In order to receive a part of that fund, a county had to organize an agricultural society and to raise a sum equal to the amount to be received from the state. These societies were authorized to elect certain officers, "all of whom shall be practical farmers." Each society was required to send in an annual report of its activities. According to this act, the president of the state society and a delegation chosen by that body should constitute the "Board of Agriculture," which should receive and examine reports from the county organizations. Those deemed worthy were to

and farmers in North Carolina were, perhaps, less articulate than most. For example, the winners of awards for essays at the State Fair in 1857 were N. T. Sorsby—an Alabama planter, but formerly of Nash County; *Doctor* S. G. Ward; and a contestant who signed as "Edgecombe." Edmund Ruffin noted that in a group of Edgecombe County farmers "there were too many who said nothing, and who seemed too diffident, and modest, to act otherwise than as attentive listeners." *Farmer's Journal,* III (October, 1854), 195.

[14] Hillsboro *Recorder,* January 24, 1821. For an excellent account of the activities of the state society, see H. D. Wilkin, "Promotion of Agriculture in North Carolina, 1810-1860," 80-141.

[15] *American Farmer,* III (January, 1822), 339. The Board of Agriculture, after its organization in 1823, assumed responsibility for publishing the winning essays.

be published annually by the board, at state expense, and distributed throughout the state. The board was given an appropriation of $1,000 for the payment of necessary expenses and to purchase and distribute seed in the state.[16]

In pursuance of this act, a board of agriculture was organized in 1823 and Duncan Cameron, of Orange County, was chosen president of the body.[17] Annual meetings were held for several years thereafter. The principal contributions of this organization were the several volumes it published on various subjects connected with agriculture, and its distribution of agricultural literature, seeds, plants, grape vines, and silk worm eggs through the local societies.[18]

As was true in other states, the agricultural societies of North Carolina lost during the 1820's the impulse which inspired their founding and fell into decline and inactivity. Numerous explanations have been offered for this fact. Wayne C. Neely attributed their decline in the nation at large to the paucity of state aid given them, the low price of farm products after 1815, the lack of markets for farm surpluses, and the fact that the societies did not prove attractive to "working farmers."[19] In North Carolina, the decline was brought about by class cleavage, poor transportation, indifference, and the small number of "progressive farmers."[20] To these may be added the undoubted disappointment of members who found no solution to their problems in these societies, the fact that the programs and conduct of meetings were usually pitched on a plane which not only failed to stimulate but actually repelled the "average" farmer from participation, distraction of leaders by political issues, and the lure of the West which became increasingly strong in the 1820's and 1830's.[21]

Numerous announcements of the organization of agricultural societies appeared in the farm journals but nothing further was heard from them. The *Farmer's Register* ceased publishing their constitutions, believing, as the editor said, that it were best for the societies "not to

[16] *Laws of N. C.* (1822), 18-20. This act was amended in 1824 to authorize two more years in which to qualify for state aid to those counties which had not yet formed societies. *Ibid.* (1824), 7-8. For contemporary comment on this act, see Salisbury *Western Carolinian,* May 4, 1824.

[17] *American Farmer,* V (December, 1823), 293.

[18] *Ibid.,* X (January, 1829), 336; *ibid.,* XI (January, 1830), 346. Included in these volumes were reports of the geological survey made by Denison Olmsted and Elisha Mitchell and the essays selected by the board for publication.

[19] W. C. Neely, *The Agricultural Fair,* 70-71.

[20] H. D. Wilkin, "Promotion of Agriculture in North Carolina, 1810-1860," 28-29.

[21] A contributor to the *American Farmer* in 1820 "lamented" the fact "that the proceedings of many of these associations, should be marked with so little of that simplicity and economy, which became *the Americun* [*sic*] *Farmer.*" *American Farmer,* II (April, 1820), 16. Edmund Ruffin chided the county societies on their small accomplishments and urged them to "take the rare course of drawing out the talents of . . . members." *Farmer's Register,* III (January, 1836), 575.

publish their *sayings* until they can accompany them with something of their doings. . . ."[22]

In 1823, Governor Gabriel Holmes expressed regret that more counties had not participated in the aid offered by the previous legislature by forming agricultural societies. Governor Holmes recommended that the legislature present each society with copies of the *American Farmer* and that future legislatures do their "mite" towards encouraging agricultural reform. To those discouraged with the progress thus far, he said: "Such is the rooted strength of old customs and modes of farming, particularly amongst us, that we must expect improvement to be gradual. . . ."[23]

Though they did not disappear entirely the county societies became less numerous in the late 1820's and 1830's and did not revive until after 1845. The state society disappeared so completely that the similar body formed in 1852 was considered a new organization rather than a revival of an old one.

Changing conditions brought about a revival of interest in agricultural societies in the 1840's both in North Carolina and in the nation. Wayne C. Neely attributed the growth of this new interest to the general expansion of agriculture; the publication of Justus Liebig's work on the chemistry of soils in 1840; improved transportation, communication, and marketing facilities; improved livestock; improved farm implements; the repeal of the Corn Laws in Englind; the potato famine in Ireland; a rise in farm prices after 1845; increased travel; an increase in reading by the general public; the influence of the first state fair in New York in 1841; and, the organization of the National Agricultural Society in 1841.[24] Other contributing factors in the case of North Carolina were the fact that the tide of emigration from the state passed its peak somewhere in the decade of the thirties; the growing influence of Edmund Ruffin's work as a planter and an agricultural reformer; the rising trend toward general progress in the state after 1835; and, recognition of the greater necessity for reform.[25]

Several County societies became active after 1845, and by the end of

[22] *Farmer's Register,* III (January, 1836), 575.

[23] *Niles' Weekly Register,* XXV (November 29, 1823), 200.

[24] W. C. Neely, *The Agricultural Fair,* 72-79. Solon Robinson, with a group of "friends of agriculture," organized the National Agricultural Society. James M. Garnett, of Virginia, became first president of the body, and Edmund Deberry, of North Carolina, was elected one of the vice-presidents. *Niles' Weekly Register,* LXI (December 11, 1841), 226; (January 1, 1842), 273. For an excellent appraisal of the state of agriculture in the United States at this time, see Robinson's remarks to Jesse Buel, editor of the *Albany Cultivator,* concerning the organization of the national society. H. A. Kellar (ed.), *Solon Robinson,* I, 87-90.

[25] Writing of the general aspects of agriculture in North Carolina, an observer said in 1834: "Very little has yet been done towards the improvement of our plantations. Manure is preserved in but small quantities, and few aim at any thing higher than the *two-shift system.*" *Farmer's Register,* I (March, 1834), 614-615.

the decade the movement was underway in earnest.[26] In 1852, the state provided a subsidy of $50 annually for each county society provided it raised an equal amount and submitted annual reports of its financial operations. This money was for the purpose of paying necessary expenses and buying awards for winners of contests sponsored by the society. Each society was also required to publish in at least one newspaper a report of its experiments and improvements.[27]

The form of organization, constitutions, by-laws, officers, programs, membership, and objectives of these societies were similar to those of the earlier period. The same deficiencies and tendencies may also be noted. Again, the societies failed to attract membership from the mass of farmers. One editor was led to remark in this connection that most farmers were ignorant of accomplishments being made in agriculture. "And this is not all," he wrote, "—they manifest a determination to remain ignorant. . . ."[28]

In its conduct of meetings, the Scotland Neck society was unusual. It used the "conversational" rather than the "speaking" method at its meetings. No one was required to rise from his seat in addressing the president and "no display in speech making" was attempted. This procedure encouraged a wider participation in the discussions; and, according to an observer, "by far the most interesting and instructive remarks are made by persons, who, if required to place themselves in a speaking attitude would not say one word."[29]

The Bear Marsh Agricultural Club also operated in a novel manner. Meetings were held monthly at the home of a member, the selection being made by ballot, where, upon arrival, the group would make a tour of the plantation. The host conducted this tour and informed his guests as to the manner of cultivation, quantity and quality of manure applied, and other aspects of his operations. After dinner was served, the group was called to order by the president for the purpose of debating such questions as arose. In the meantime, two members appointed by the group prepared a criticism of the host's farm which was discussed by the club and either accepted or rejected by vote. Of this organization, a member wrote: "Since the organization of our Club, our interest in agriculture has regularly grown greater, and our manner of cultivation more improving and successful. We do more work, and better work, and enjoy greater returns than formerly."[30]

The society-sponsored county fair assumed larger proportions, and

[26] The societies in Granville, Orange, Mecklenburg, and Buncombe counties were said, in 1845, to be "marching forward in the work of improvement." *North Carolina Farmer,* I (June, 1845), 19.

[27] *Laws of N. C.* (1852), 4-7. See article on this act in Hillsboro *Recorder,* March 9, 1853.

[28] *Farmer's Advocate,* III (January, 1842), 235.

[29] *Farmer's Journal,* II (September, 1853), 211.

[30] *North Carolina Planter,* II (December, 1859), 355-356.

was, no doubt, the most effective means of attracting public interest and participation in the society's program. Mecklenburg held a county fair each year after 1842.[31] Several other counties did likewise. The regional fairs of the North Carolina-Virginia Agricultural Society at Petersburg, and of the Seaboard Agricultural Society at Norfolk, attracted a great deal of interest in North Carolina towards the end of the period.

Principally because of the initiative taken by Doctor John F. Tompkins, editor of the *Farmer's Journal,* a meeting was held in Raleigh in October, 1852, and the North Carolina Agricultural Society organized. No reference was made to the previous organization by this name. "Let it no longer be said," Tompkins wrote, "that the farmers of North Carolina are blind to their interests, and that they have no spirit of pride as regards the advancement of farming among them."[32] Though only a few delegates appeared at the appointed time, much to Tompkins' "shame and mortification," this organizational meeting was attended by representatives from eighteen counties, scattered in all sections of the state. John S. Dancy, of Edgecombe County, was elected president; plans were made to conduct a state fair in the next year; and a memorial to the legislature was drawn in which an appropriation was requested from the state to carry out the society's program.[33]

The memorial declared that agriculture in North Carolina was "in a languishing condition," "too much neglected by men of science," and that the state, though possessing unsurpassed climate and soil, occupied a "humiliating position in the rear of her sisters, in Agricultural and Industrial improvement."[34] The memorialists charged the state with neglect of her farming interests and the interests of her laboring classes, with being too much absorbed in federal politics, and with failure to encourage "competition and skill in the development of her great agricultural resources."[35] The legislature, however, failed to take any action on the memorial.

From its lowly beginning, the state society quickly sprang to life under the direction and leadership of such men as Doctor Tompkins, John S. Dancy, Judge Thomas Ruffin, Doctor William R. Holt, Kenneth

[31] For accounts of the county fair in ante-bellum North Carolina, see H. D. Wilkin, "Promotion of Agriculture in North Carolina, 1810-1860," 51-79; Wade Phillips Young, "A History of Agricultural Education in North Carolina," 36-55, unpublished doctoral dissertation at the University of North Carolina (1934); G. G. Johnson, *Ante-Bellum North Carolina,* 106-109. Governor John W. Ellis, in his message to the legislature in 1860, said that the county fairs had done more to stimulate interest in agricultural reform in the state than any one thing. *Legislative Documents* (1860-1861), 15. For an excellent coverage of agricultural fairs in the nation, see W. C. Neely, *The Agricultural Fair, passim.*

[32] *Farmer's Journal,* I (July, 1852), 112, 114.

[33] *Farmer's Journal,* I (November, 1852), 243-244, 246-247, 250. The society was incorporated by the state in December, 1852. *Laws of N. C.* (1852), 3-4.

[34] Raleigh *Register,* November 3, 1852.

[35] *Ibid.*

Rayner, Paul C. Cameron, Henry K. Burgwyn, and Richard H. Smith. In the first year, the membership increased from nineteen to four hundred. "The meetings instead of being thinly attended, as at first, seem to interest all," Doctor Tompkins wrote.[36] By January, 1855, the society had more than one thousand members.[37] Dancy was followed in the presidency by Richard H. Smith, of Halifax County, who in turn was succeeded by Judge Thomas Ruffin in 1854. Judge Ruffin was re-elected annually until 1859 when he was permitted to step down and Doctor William R. Holt, of Davidson County, was elected. These men were outstanding leaders in the economic. political, and social life of the state. As will be seen, they also ranked with the best as enlightened, practical farmers. Assisted by executive committees composed of equally progressive minded individuals, these presidents directed the state society through the remainder of the period—each year of which was marked by an increase in membership and by real accomplishments in the interest of agriculture.

The state society encouraged a generous support of agricultural periodicals, particularly the *Farmer's Journal,* which was adopted as the official organ of the society and to which five hundred subscriptions were taken for distribution in the state.[38] Though no other paper was given such favored backing after the failure of the *Farmer's Journal* in December, 1854, the society continued to urge farmers to subscribe to farm papers.

Under Dancy's presidency, the society sponsored a "corn sweepstakes" in which prizes were offered for the largest per acre yield of corn. In order to qualify for these prizes, the contestants were required to submit a complete account of their methods, types of soil, manures used, preparation of soil, seed used, times and modes of culture, weather conditions, and of any other factors influencing production. A disinterested person measured the field and the product for the above report. The winners of this contest were J. W. Woodfin, of Buncombe, Joseph Cathey, of Haywood, and Sidney Weller, of Halifax counties, who produced on one acre 109, 101, and 70$\frac{3}{5}$ bushels respectively.[39]

The state society also offered sizable annual premiums for essays on a variety of agricultural subjects. The best of these were published in farm journals, newspapers, or in the transactions of the society.[40] As a part of the reform program, farmers were encouraged to preserve such articles to which they might refer for advice and direction in solving their own problems. There is much evidence in the farm papers that many of the new practices and ideas in agriculture, as well as new

[36] *Farmer's Journal,* II (November, 1853), 242.
[37] *Carolina Cultivator,* I (March, 1855), 2.
[38] *Farmer's Journal,* II (November, 1853), 242.
[39] Raleigh *Register,* March 4, 1854.
[40] See for example, *Transactions of the North Carolina Agricultural Society* (1857), 34-92.

implements, seed varieties, and breeds of livestock were popularized in this manner.

The most popular and, perhaps, influential activity of the state society was the state fair, which it conducted annually at Raleigh after 1852. This fair was held in October each year. In attraction to the people and scope of display, the fair grew steadily until 1860 by which time it was truly a state-wide institution, offering exhibits by and to people from all over the state.

Beginning with a cash premium list of only $524 in 1853, the amounts given in cash awards and other prizes increased yearly.[41] The state gave its first positive support to the fair, in 1855, when the legislature appropriated $1,500 to match a similar sum raised by the society which was to be given as premiums. This was to be an annual appropriation "to advance the interest of agriculture and manufactures."[42] Although apathetic at first, the business people and citizens of Raleigh contributed a great deal towards the success of the fairs.[43] The railroads leading into Raleigh, except the Petersburg Railway, gave tremendous impetus to attendance and competition by carrying exhibits to and from Raleigh free, and passengers for only half-fare.[44]

Several state newspapers rendered an important service in the movement toward agricultural reform by conducting columns devoted to agricultural subjects. The Raleigh *Star,* as previously noted, first entered this field. Notable among the papers conducting similar columns were the Hillsboro *Recorder,* Tarboro *Southerner,* Salisbury *Carolina Watchman,* and the Salisbury *Western Carolinian.*[45] Although used to some extent for editorial remarks and as an open forum for local contributors, this space was most often used to carry reprints from out-of-state farm journals or the addresses and essays presented to agricultural societies.

So few articles were by local authors that "A Young Orange Farmer" was led to protest as follows:

> I read the part of your paper that is devoted to agriculture with pleasure; and often ask, why is it that I see almost every piece on the pursuit of agriculture taken from some other paper, when we have so many good farmers in our own county? All of them it seems, have laid down their pens.[46]

[41] For an account of the first fair and a listing of premium winners, see *Farmer's Journal,* II (November, 1853), 225-242.

[42] *Laws of N. C.* (1853-1854), 3.

[43] J. G. deR. Hamilton (ed.), *Papers of Thomas Ruffin,* II, 493, 496.

[44] *Carolina Cultivator,* II (November, 1856), 273.

[45] In 1832, there were twenty-five newspapers in North Carolina. Most of them were very poorly supported and many were short-lived, because, as Professor Hamilton has put it, "in truth North Carolina had no reading public, not even a public that read newspapers." J. G. deR. Hamilton, *Party Politics in North Carolina, 1835-1860,* 24.

[46] Hillsboro *Recorder,* April 9, 1846. For accounts of the influence of the

Of more significance to the mass of farmers, perhaps, than the newspapers, were the almanacs, which were widely circulated and read in the state. These publications carried some technical information about farming which furthered the cause of reform.[47]

This period witnessed the launching of many journals devoted almost exclusively to farm subjects. Unfortunately, the enthusiasm that prompted many of these ventures waned with their failure to attract sufficient backing; and, like many of the agricultural societies, they were short-lived. Of the seven purely agricultural journals published in the state between 1838 and 1860, none lasted longer than four and one-half years, and at the latter date, only two were being published. This experience was not peculiar to North Carolina alone but was typical of what happened throughout the nation. Notwithstanding the obvious failure to attain a degree of permanency, these papers, and others circulated in the state, played an important part in furthering the cause of agricultural reform by keeping their readers informed of current thought and practice in agriculture.[48]

Of the out-of-state farm journals, the most popular and influential in North Carolina were the *American Farmer,* published at Baltimore (1819-1839); the *Southern Agriculturalist,* published at Charleston (1828-1846); and, the *Farmer's Register,* published at Shellbanks and Petersburg (1833-1843). Though designed as regional rather than as state papers, not one of these continued uninterrupted publication to 1860! In 1836, at a time when not a single farm paper was being published in North Carolina, there were only one hundred and ten subscribers in the entire state to Edmund Ruffin's *Farmer's Register.* Thirty of those lived west of Raleigh. New Hanover County, which contained the largest city in the state, had only one subscriber.[49]

*The Farmer's Advocate and Miscellaneous Reporter,* published at Jamestown (1838-1842), with John Sherwood as editor, was North Carolina's first farm journal. Thomas J. Lemay edited and published the *North Carolina Farmer* at Raleigh (1845-1849). The *Farmer's Journal,* edited by Doctor John F. Tompkins, was published at Bath and Raleigh (1852-1854). The *Carolina Cultivator* was published at Raleigh (1855-1857), with William D. Cooke and Benjamin S. Hedrick as editors. Thomas J. Lemay again entered the field with the publication at Raleigh of the *Arator* (1855-1857). The *North Carolina Planter* was

newspaper on ante-bellum North Carolina agriculture, see H. D. Wilkin, "Promotion of Agriculture in North Carolina, 1810-1860," 142-200 passim; W. P. Young, "A History of Agricultural Education in North Carolina," 412-421.

[47] For an account of those published in the state in this period, see W. P. Young, "A History of Agricultural Education in North Carolina," 409-412.

[48] One authority has ascribed to the farm journals the superior role in the reform movement of the period. A. L. Demaree, *The American Agricultural Press 1819-1860,* 12-13.

[49] *Farmer's Register,* III (April, 1836), 786.

published at Raleigh (1858-1861). On the staff of editors of this paper were James M. Jordan, John W. Woodfin, S. W. Westbrook, and W. H. Hamilton. The *Edgecombe Farm-Journal,* published and edited at Tarboro by William B. Smith (1860-1861), completes the list.

Though several of these periodicals, particularly the *Farmer's Journal,* possessed real merit, not a single one was adequately supported. John S. Dancy, president of the state agricultural society, "grieved" with Doctor Tompkins that his paper was not better patronized by the farmers of the state, who, he alleged, were "handsomely" supporting several out-of-state journals. Dancy attributed this failure to a want of state pride.[50] Doctor Tompkins said many of the farmers were so prejudiced and opposed to innovations that they considered everything projected for their benefit "a 'Yankee humbug,' and thus they repulse us, thinking our intention is rather to injure than benefit them."[51] In 1859, the editor of the *North Carolina Planter* spoke of his publication as a "payingless enterprize" and exclaimed: *"We have less than eight hundred subscribers!"*[52] At that time, there were over seventy-five thousand farmers in the state and his was the only farm journal! To make bad matters worse, several of the editors were led to complain that many subscribers would not pay their bills. "They are," wrote one. "in many cases, so long withheld, that we are left to consider them as *indefinitely postponed."*[53] Perhaps no better explanation can be offered for the above conditions than to concur with Professor Hamilton in saying: "At the period of which we speak, the state as a whole was ignorant, shockingly so, even for that time. . . ."[54]

Doctor Tompkins was undoubtedly the ablest of the editors of the farm journals published in the state. As stated by him the *Farmer's Journal* was "devoted exclusively to the setting forth of the various popular improvements in Agriculture, Horticulture, and the household arts."[55] Because of his enthusiasm for reform and energy in promoting it, the spirit of inquiry then stirring in the state was given increased impetus. The result was that many farmers were led into experimentation, competition, and more rapid improvement. "Our good old State," he wrote, "is far behind the age in Agricultural, as well as every other improvement. As a people, we are greatly wanting in State pride, which

[50] *Farmer's Journal,* II (January, 1854), 308.

[51] *Ibid.,* II (October, 1853), 208. "A. O. G." wrote the editor of the *North Carolina Farmer* that those who did not approve of his paper might be classified into three categories: those who had neither time to read it nor money to pay for it; those who were opposed to "book farming" and *"in fact,* to all *innovations, inventions,* and *improvements of every kind";* and, those who considered his publication a "money trap." *North Carolina Farmer,* I (February, 1846), 211.

[52] *North Carolina Planter,* II (September, 1859), 273-274.

[53] *North Carolina Farmer,* II (January, 1847), 192; see also, *Farmer's Advocate,* III (September, 1842), 369.

[54] J. G. deR. Hamilton, *Party Politics in North Carolina, 1835-1860,* 26.

[55] *Farmer's Journal,* I (April, 1852), 1.

is highly important to place us in that position which we ought to occupy."[56]

All the journals had some enthusiastic supporters and it was mainly through the efforts of supporters that the subscription lists were extended. "F. L. B.," of Edgecombe County, sent in a list of fifty new subscribers to the *Southern Cultivator* and said:

> Agriculture, in this county, has improved 200 per cent on an average, and parts of it 300 per cent within the last six years. And I assure you I cannot attribute it to any other cause than the improved mode of Farming, which the *Cultivator* and similar journals advocate.[57]

J. G. Anson, a "young" planter of Anson County, wrote as follows: "The publication of your journal has excited a spirit of inquiry among the agriculturists, and they begin to interchange opinions, establish theories, institute experiments, and deduce practical inferences. . . ."[58] Though inclined towards optimism himself, Doctor Tompkins could not have wished for a more generous reception for his *Farmer's Journal* than "Mentor," of Hertford County, gave in the following:

> There is indeed a spirit of inquiry spread among our farmers which did not exist before *The Farmer's Journal* came among us. Go into what part of the county you will, it is now evident that the farmers are up and doing; you may see large compost heaps being made, ditching being done, swamp mud thrown up, woods mould collected, and the farmers are heard to say that they never were so busy before. . . . Our crops next year, with fair seasons, will speak trumpet-tongued in favor of heavy manuring, deep plowing and thorough drainage. Our farmers here feel that there is a brighter day coming.[59]

Generally, the editors of farm journals sought to emphasize and urge acceptance by their readers of whatever was considered the best thought and practice by agricultural leaders at the moment. By selection of articles from the pens of Edmund Ruffin, Solon Robinson, Jesse Buel, Daniel Lee, and other national leaders, as well as by reproducing the proceedings of English and Scotch agricultural societies, effort was made to keep the reader abreast of the times. Not even Cato, Columello, Varro, and other ancients were overlooked in the search for the fundamental truths about farming. The constancy of the editors in advocating reforms contributed immeasurably toward making farming more efficient, intelligent, enjoyable, and profitable.

Among the things most often advocated were deep plowing, horizontal plowing, thoroughness of cultivation, keeping the soil in "good heart" by scientific rotation of crops, planned orderliness on the farm,

[56] *Farmer's Journal*, I (April, 1852), 1.
[57] *Southern Cultivator*, VIII (December, 1850), 185.
[58] *Southern Agriculturalist*, II (May, 1829), 222-223.
[59] *Farmer's Journal*, I (September, 1852), 161.

record keeping, improvement of livestock, use and care of proper implements, selection of only the best seed for planting, agricultural education, and internal improvements. In all these things, as Albert L. Demaree has said, they worked hand in hand with the agricultural societies for "their objectives were identical," and "the benefits of co-operation were mutual."[60] Generally, too, the editors of the farm journals published in North Carolina studiously avoided politics and controversial questions not directly connected with farming. This was the case with most farm papers published in the period.[61] Further consideration of reforms emphasized by the agricultural press will be given later in the chapter.

In the late ante-bellum period, there was a great increase in the number of books published on agricultural subjects. These were advertised in the farm papers and sold in the book stores of the state. Richard Lamb Allen's *The American Farm Book; or Compend of American Agriculture. . .*, published in 1850, is a good illustration of this type of literature. This volume was designed as a manual on agriculture and dedicated to "The Farmers & Planters of the United States." As revealed in its subtitle, it was "a Practical Treatise on Soils, Manures, Draining, Irrigation, Grasses, Grain, Roots, Fruits, Cotton, Tobacco, Sugar Cane, Rice, and Every Staple Product of the United States with the Best Methods of Planting, Cultivating, and Preparation for Market." This volume and Miner's, *Bee-Keepers Manual;* Brown's, *American Poultry Yard;* Allen's, *Domestic Animals;* Johnston's, *Lectures on Practical Agriculture;* Buist's, *Family Kitchen Gardener;* Brown's, *American Bird Fancier;* Dana's *Prize Essays on Manures;* and, Rich's, *American Architect,* all of which were published by C. M. Saxton, of New York City, were advertised as premiums by the *American Agriculturist,* and a copy of one of them was given to each new subscriber. It goes without saying that the widespread prejudice against "book farming" kept down the sale of these volumes.

To rid the farmer of his prejudice became an important feature of the reform program. The editor of one state newspaper raised the issue with his readers: "Can Farming be an Intellectual Pursuit?" In partial answer to this query, he strongly advocated a "spirit of inquiry" concerning every activity of the farm. "The time is fast coming," he wrote. "when our farmers must look at more intently, and comprehend more clearly, the principles of good husbandry."[62] Edmund Ruffin admitted he was a "book farmer" and wrote an essay in the "book farmer's" defense.[63] There is considerable evidence that many farmers were

[60] A. L. Demaree, *The American Agricultural Press, 1819-1860,* 39-43, 69-76.
[61] *Ibid.,* 79.
[62] Salisbury *Carolina Watchman,* May 4, 1848.
[63] Edmund Ruffin, *Essays and Notes on Agriculture,* 278-285.

brought into an appreciation of the value of "book learning" through reading farm journals. A Beaufort County farmer wrote:

Three years ago, I thought book farming a humbug—that the mechanical condition of the soil was all that required a farmer's attention; and the men that talked oxygen, hydrogen, and such hard words to me, I thought was making fun, because I supposed that none but educated men, who had devoted their minds to the study of chemistry, could understand such matters; and that farmers had no occasion to know anything of the subject.[64]

This man claimed that by studying the soil and applying the lessons learned from books he had increased the yield of corn from a thirty acre field by ninety barrels in one year.[65]

Of great significance, too, is the fact that agriculture became a matter of concern and field of activity by government, both national and state. Notice has been taken of the passage of the national patent law which gave encouragement and protection to those engaged in improving farm implements and machinery. In 1836, the Patent Office, in addition to its primary function, was given the responsibility of looking after the agricultural interests of the national government. Given only a small appropriation and otherwise limited in its scope of activity, this agency was criticized by the farm press almost from the beginning for "neglect, political corruption, and incompetency."[66] Its chief accomplishments to 1860 were in the distribution of seeds and plants and the publication of annual reports on various aspects of agriculture.[67] From less than twenty pages in 1843, these reports steadily increased in variety and scope uptil they filled a large volume in the annual report of the office during the fifties.

Henry King Burgwyn, of Northampton County, was the author of an article on the subject: "Improvement of Worn-Out Lands By the Use of Peas and Clover," which appeared in the Patent Office report for 1849.[68] The editor of the *Southern Cultivator* reprinted the article by Burgwyn with the following comment:

If the Agricultural Report (of which one branch of Congress has ordered 100,000 copies to be printed, and the other will, doubtless, order half as many more) contained no other information than the above paper,

[64] *Farmer's Journal,* I (April, 1852), 2.

[65] *Ibid.*

[66] A. L. Demaree, *The American Agricultural Press, 1819-1860,* 56.

[67] The Patent Office began distributing seed in 1840, when it sent 30,000 packages to farmers throughout the United States. In 1847, the number of packages distributed had risen to more than 60,000, of which, about one-half had been contributed by the French Minister of Agriculture. In 1852 and 1854 the authority of this office to purchase seed for distribution was increased. Cuttings were also included. Alfred Charles True, *A History of Agricultural Experimentation and Research in the United States, 1607-1925, Including a History of the United States Department of Agriculture,* 25-28, cited hereinafter as *History of Agricultural Experimentation.*

[68] *Report of the Commissioner of Patents (1849),* 400-402.

from an eminently practical man, on the improvement of "Worn-out Lands" we should regard the money as well expended.[69]

In addition to articles like the above, the commissioner's report contained crop estimates, shipping information, weather information, scientific data related to agriculture; and, what was of more value to farmers, the replies of practical farmers from every state to a circular sent out by the Patent Office. This circular contained questions about practically every crop and all kinds of livestock grown in the nation. This part of the report provided farmers with a storehouse of information upon which to base experiments in new methods and in new crops.[70] Toward the end of the fifties, over a quarter of a million copies of the commissioner's report were printed annually. These were made available for distribution to either individuals or agricultural societies by application through senators and representatives.

Beginning with the census of 1840, the Department of the Interior collected and published a great mass of agricultural statistics in connection with the decennial census taking which attracted considerable interest and proved of value to those seeking agricultural reform.[71] By 1860, this report alone filled a large volume of the census and included not only statistics of production, but also evaluations of trends, the story of the evolution of farm implements, and a grouping of the states in accordance with their geographic, climatic, and agricultural similarities. This last facilitated sectional comparisons—an activity that engaged the attention of many in that day. Some of the figures contained in these reports were so startling in their revelation that their accuracy was challenged. "X," for example, was amazed to learn from the census of 1840 that Tennessee was producing almost twice as much tobacco as Virginia and declared the census "grossly" in error.[72] "W" came out just as strongly in defense of the accuracy of that census.[73] Though less frequent, somewhat similar discussions arose after the publication of each new census.

In this period, too, the government of North Carolina made an immeasurable contribution to agricultural reform by assuming a broader concept of its functions and responsibilities. This change did not come easily, quickly, or all at once, but came as the gradual fruition of many

[69] *Southern Cultivator,* VIII (May, 1850), 65.

[70] Thomas P. Devereux, of Halifax, W. J. Bingham of Orange, R. H. Bennet of Iredell, Joshua Harris of Cabarrus, and, G. S. Sullivan of Lincoln County made replies to these circulars.

[71] For example, one editor, using material concerning agricultural trends in England during the previous fifteen years contained in the Census of 1840, showed how those trends had directly influenced agricultural development in the United States. *Niles' Weekly Register,* LXIII (September 10, 1842), 25-28.

[72] *Farmer's Register,* IX (July, 1841), 437.

[73] *Ibid.,* IX (August, 1841), 490-492.

years of effort by political and social leaders, chief among whom was Archibald D. Murphey.[74]

Contemplated within the program of these reformers were projects for appraising the resources of the state, establishing a system of internal improvements, establishing a public school system, draining state-owned swamp lands, and a revamping of the political order in the interest of greater democracy and a more equal balance in political representation. Each of these projects had tremendous import to the agricultural interests of the state; and, by 1860, substantial progress had been made toward the accomplishment of each.

In 1823, a geological and mineralogical survey of the state was authorized by the legislature. This survey was to be made under the direction of the board of agriculture by "some person of competent skill and science" selected by that body.[75] Denison Olmsted, of the faculty of the University of North Carolina, was employed to make the survey. Professor Olmsted was assisted in the work by Elisha Mitchell of the same institution. Their findings were published "for the benefit of the public."[76] This was the first survey of this nature made by any state. The report dealt chiefly with the location and distribution of deposits of gold, coal, and building stone, and an appraisal of the suitability of the limestones and marl found in the state for use as fertilizers.[77] This re-report, taken with those being submitted at the same time by Hamilton Fulton, state engineer, provided reformers a body of information about the state that was as broad in scope and as scientific in its appraisal as that available on any state. Considerable question may be raised, however, as to how intelligently that information was used.

Archibald D. Murphey, in his famous "Memoir" on internal improvements in 1819, recommended that the state undertake the drainage of swamps and marsh lands for the two-fold purpose of safeguarding the health of the inhabitants and reclaiming and subjecting to agriculture "a large portion of the richest lands in the state."[78] He directed attention to the experiments in drainage that had already been made at Lakes Phelps and Mattamuskeet, and took the position that such proj-

[74] Governor David L. Swain, in his message to the legislature in 1833, pointed out that ordinarily the expenses of that body had exceeded all the other costs of government in the state, and expressed "a doubt whether its legitimate functions have been performed." He urged upon the legislature "the propriety of entering upon a system of legislation required by the wants of your constituents, commensurate with their resources, and worthy the confidence which they repose in your ability to administer their public affairs." *Legislative Documents* (1833), 4.

[75] *Laws of N. C.* (1823), 17.

[76] *American Farmer,* VI (April, 1824), 14; A. C. True, *History of Agricultural Experimentation,* 12-13.

[77] J. A. C. Chandler, and others (eds.), *The South in the Building of the Nation,* V, 553, 556-557.

[78] *Report on Internal Improvements in North Carolina, 1819-1820,* 37.

ects were beyond the means of individuals and could only be accomplished by state action.[79]

During the twenties, the legislature directed the state board of internal improvements to make surveys of the swamp lands belonging to the state, and to obtain data as to the practicability of their drainage.[80] Very little was accomplished prior to 1837, however, beyond the procurement of survey data. In that year, the state literary board was given control of all swamp lands, belonging to the state, which were to be used as a source of funds for the establishment of public schools. The board was directed to adopt ways and means of surveying, draining, and selling those lands. As a starter, the board was given $200,000 from the funds distributed to the state by the United States.[81] By 1840 the literary board had expended $130,000 and drained between 70,000 and 80,000 acres of swamp lands lying principally in Hyde, Washington, and Tyrrell counties. When offered for sale, some of this land sold for as much as $40 an acre.[82] Times were hard, however, and the bids for this land were generally so low that the state ceased trying to sell it.[83] In 1847, the literary board in accordance with an act of the legislature again offered these lands for sale.[84] Because of disputes arising as to titles and of the difficulty of access to the lands, the public showed only moderate interest in the sale. In 1854, Governor David S. Reid spoke of the swamp lands as an uncertain source of state revenue, and suggested that further legislation was necessary in order to safeguard the public's interest in them.[85] Professor Ebenezer Emmons attempted to attract a more active interest in these lands by praising their qualities for agricultural purposes, and contending that the soil was "fully equal in productiveness to the prairie lands of Illinois. . . ."[86] In all, to 1856, the state spent $191,364 on drainage projects. The state's policy of holding the price of the reclaimed land high, however, reduced

[79] *Report on Internal Improvements in North Carolina, 1819-1820,* 37. It was estimated in 1827 that the state owned about one and one-half million acres of swamp lands and that nearly a million more were owned by individuals. Tarboro *Press,* November 17, 1827. As to the value of such lands, the state engineer pointed to a reclaimed parcel of land in Beaufort County, on which from thirteen to sixteen barrels of corn per acre were being produced. This same land, prior to drainage, was offered for twenty-five cents per acre without any takers.

[80] *Laws of N. C.* (1826), 21-22.

[81] *Ibid.* (1836-1837), 131-135. Provision was made in this act that individuals whose lands were improved by the state projects should pay a proportionate part of the costs.

[82] Raleigh *Register,* October 30, 1840.

[83] Governor John Motley Morehead to the North Carolina General Assembly, 1844, *Legislative Documents* (1844-1845), 14-15.

[84] *Laws of N. C.* (1836-1847), 248; Raleigh *Register,* April 27, 1847.

[85] *Legislative Documents* (1854-1855), 5-6.

[86] Ebenezer Emmons, *The Swamp Lands of North Carolina,* 59. See also Ebenezer Emmons, *Report of the North Carolina Geological Survey. Agriculture of the Eastern Counties. . . . , passim,* cited hereinafter as *Report of the North Carolina Geological Survey.*

the public's interest in these lands. In 1860, most of the swamps remained undrained and still in the possession of the state.

This period, too, witnessed developments in the realm of education, both general and theoretical, that were of major significance to agriculture. In his message to the legislature in 1823, Governor Gabriel Holmes recommended the purchase of a farm near the University "to be put under the care of a scientific and practical farmer," on which experiments might be conducted, under the direction of the professor of chemistry, in manuring and cultivation of crops so that the students might "*see* and learn the usefulness and beauty of husbandry."[87] "The person having charge of the farm," he said, "should understand the mechanism and use of the most improved implements of husbandry, and also be well informed in the different departments of domestic economy."[88] Governor John M. Morehead made a similar recommendation to the legislature in 1844, and also advocated the establishment of an "agricultural professorship" at the University; or, if considered more advisable, the establishment of a school where agriculture might be taught as a science.[89] In 1853, a professorship of agricultural chemistry was established in the University and Benjamin Sherwood Hedrick became the first occupant of the position.[90] To a contributor in the Fayetteville *Carolinian,* that was not enough. ". . . we want a school," he wrote, "where the practice as well as the theory of Agriculture may be taught, where theory can be verified by practice, and where the results of that practice may be scattered among the farming community throughout our state."[91] The state, however, did not see fit to take further action in this direction.

Several attempts were made by private agencies to establish agricultural schools. The Episcopal Church founded a school at Valle Crucis in Ashe County in 1845 to instruct young men "in the sciences auxiliary to agriculture, illustrated by practical farming."[92] In 1853, announcement was made that an agricultural school would begin at Bath the following year. Connected with this school was a thirty acre experimental

[87] *Niles' Weekly Register,* XXV (November 29, 1823), 200-201.

[88] *Ibid.*

[89] *Legislative Documents* (1844-1845), 10-11. For an account of the various proposals for the teaching of agriculture at the University between 1795 and 1852, see W. P. Young, "A History of Agricultural Education in North Carolina," 24-36.

[90] K. P. Battle, *History of the University of North Carolina,* I, 642-644. While in that connection, Professor Hedrick also served as agricultural editor of the *Carolina Cultivator* until his resignation in November, 1856 "for reasons," as he stated, "known to the public." *Carolina Cultivator,* II (November, 1856), 273.

[91] *North Carolina Planter,* I (November, 1858), 340.

[92] *North Carolina Farmer,* I (June, 1845), 21. In commenting on this school, the editor appealed to "capable and aspiring young men to avoid the crowded avenues of those professions (law, medicine, etc.), and find usefulness, honor and happiness in the quiet and shady walks of agriculture." Albert L. Demaree has said that the agricultural press of the United States in this period was more persistent in its advocacy of agricultural education than in any other measure of reform. A. L. Demaree, *The American Agricultural Press, 1819-1860,* 51.

farm, on which, the students would not be required to work, "but only to watch the various experiments while being made."[93] J. M. Lovejoy's Academy, in Raleigh, put a course in its program entitled, "Agricultural, Elementary and Experimental Chemistry" in 1853 and employed Doctor John F. Tompkins to teach it.[94] "J. C.," in 1856, came out strongly in favor of a state-supported agricultural school for each congressional district.[95] A contributor to the Fayetteville *Carolinian,* in 1858, protested against the interest being shown by the public in military rather than in agricultural schools "which might be made as fully scientific as, and far more advantageous to the majority of our youth. . . ."[96] In spite of the efforts made, however, North Carolina did not have an agricultural school of any distinction in 1860.

The lack of a public school program contributed to the retardation of agricultural reform. According to a survey made in 1829 and reported in the Raleigh *Register,* of one hundred and fourteen families living in Wake County only forty-nine possessed Bibles, and, in twelve, not a single individual could read. In another section of the county, ten families were found in which not one of the adult members could read.[97] The importance of a public school program to agricultural progress was recognized by the reformers.[98] "By the Way," of Onslow County, wrote: "General intelligence is as essential to successful agriculture as it is to good legislation. The friends of agriculture must first become the friends of general education. . . ."[99] In this field, outstanding progress was made between 1840 and 1860. While the population of the state was increasing fourteen per cent between 1850 and 1860 the enrollment in the public schools increased thirty-six per cent. The percentage of illiterate voters decreased in the same decade from 29.2 to 23.1.[100] The importance of this development to agricultural reform can hardly be over-stressed.

Several indications began to appear that the reading interest of the people were broadening. In 1833, an association designed to serve as a "public" library in Chatham County was incorporated.[101] The "United Brothers Library Society" of Lenoir County was organized to promote a "circulating" library.[102] In 1834, a similar society was established

[93] *Farmer's Journal,* II (May, 1853), 49-50.
[94] *Ibid.,* II (November, 1853), 256.
[95] Salisbury *Carolina Watchman,* December 9, 1856.
[96] *North Carolina Planter,* I (November, 1858), 340.
[97] Reprinted in *Niles' Weekly Register,* XXXVI (July, 1829), 302.
[98] John Sherwood, editor of the *Farmer's Advocate,* said the lack of an adequate public school system was one of the strongest influences retarding agricultural reform in North Carolina. *Farmer's Advocate,* I (September, 1838), 36.
[99] *North Carolina Farmer,* III (August, 1847), 70.
[100] R. D. W. Connor, *Rebuilding an Ancient Commonwealth,* I, 590.
[101] *Laws of N. C.* (1833), 66-67.
[102] *Ibid.* (1831), 64.

at Providence, in Mecklenburg County, "for the purpose of promoting learning and science."[103]

In response to new developments, a second geological survey was made of the state in the 1850's. Since the earlier effort, Justus Liebig had startled the agricultural world with his volume entitled, *Chemistry in Its Application to Agriculture and Physiology.* Other foreign scientists had also made significant discoveries in the field of agricultural chemistry.[104] Frequent references to their work began to appear in the agricultural press. Furthermore, Edmund Ruffin had popularized a program of scientific farming which was not only changing the aspect of agriculture in his own state, but providing inspiration for and direction to a similar reform in North Carolina. Recognzing the changes these discoveries portended for agriculture, governors John Motley Morehead, William A. Graham, and Charles Manly proposed to the legislature in 1844, 1846, 1848 and 1850 that provision be made for a more comprehensive geological survey of the state. Such a measure was passed by the legislature in 1851. This act directed the governor to employ "some gentleman of science" to conduct an agricultural, geological, mineralogical, and botanical survey to include every county in the state. The governor and literary board were charged with supervisory responsibility.[105] Records were to be compiled and published on the nature and character of products, soils, and minerals found in the state. This act was highly praised in the state press for its liberality and farsightedness.[106]

Ebenezer Emmons, a native of New York and a professor of the University of North Carolina, was employed as state geologist and placed in immediate charge of the survey. Emmons retained this position until his death in 1863. His reports were published in five volumes between 1852 and 1860; and, by the attention given the chemistry of soils, they clearly illustrate the Liebig influence.[107] "New Hanover" recognized the connection between this survey and the program of the farm journals, and wrote the editor of the *Farmer's Journal* as follows: "Prof. Emmons is telling us where, and what, our agricultural advantages and resources are; we look to you for instruction, as to the best means of developing and using them."[108] In addition to the survey, Emmons rendered a considerable contribution to the agricultural advance of the state by his numerous articles to the state press and speeches before agricultural societies.

[103] *Ibid.* (1834), 33.

[104] M. Whitney, *Soil and Civilization,* 112-113; A. L. Demaree, *The American Agricultural Press, 1819-1860,* 66-69.

[105] *Laws of N. C.* (1850-1851), 162-163.

[106] Raleigh *Register,* January 29, 1851; Salisbury *Carolina Watchman,* January 30, 1851.

[107] J. A. C. Chandler, and others (eds.), *The South in the Building of the Nation,* V, 557; A. C. True, *History of Agricultural Experimentation,* 12-13.

[108] *Farmer's Journal,* I (May, 1852), 34.

As to specific reforms advocated by agricultural leaders in this period, none received more attention than those having to do with soil fertility. John Taylor emphasized in his *Arator* that no reform of agriculture could be effected without first enriching the soil.[109] Edmund Ruffin came to the conclusion, after prolonged experimentation, that most of the "poor" soils were acid in condition because of a deficiency in lime and that this condition could be neutralized by applications of marl.[110] Ruffin's account of his experimentation entitled, *An Essay on Calcareous Manures,* was published in pamphlet form and widely circulated in the South. His *Farmer's Register* also served as a vehicle for disseminating his ideas. The undoubted success Ruffin had in the use of marl and the persistence and sincerity of his advocacy of its use led to the development of a widespread interest in that substance. Deposits were found at numerous places in eastern North Carolina which the more enterprising farmers began to use in efforts to rejuvenate their lands.[111] Ruffin visited in the eastern part of the state in 1840, analyzed several specimens of marl, and considered them "the richest and most valuable . . . I have ever known, and the easiest to be used." But he found only a few North Carolina farmers who were using marl.[112] Ebenezer Emmons also devoted much attention to the location and analysis of marl deposits of North Carolina, and expounded on the theory and methods of their use.[113]

John Taylor and George W. Jeffreys advocated an almost indiscriminate use of marl, but Ruffin, and other agricultural leaders after 1840, pursued the Liebig concept that substances should be present in the soil in proper proportions to assure proper plant growth. Hence soil analyses to determine deficiencies in order that manures and fertilizers might be intelligently applied, became the vogue with "scientific" farmers.[114] The editor of the *Farmer's Journal* urged: "Let the soil be analyzed, and ascertain at once what is wanting . . . which the plant requires for its growth."[115] As might be expected, many were led to

[109] Reproduced in *Farmer's Register,* VIII (December, 1840), 719.

[110] *Webster's Dictionary* defines marl as "An earthy, crumbling deposit consisting chiefly of clay mixed with calcium carbonate. . . ."

[111] Marl had been used in North Carolina as early as 1818. L. C. Gray, *History of Agriculture,* II, 804-805. Denison Olmsted drew attention to the extensive marl deposits in the state in the accounts of his survey. Denison Olmsted, *Report on the Geology of North Carolina, Conducted Under the Direction of the Board of Agriculture,* Part I and II, *passim.*

[112] *Farmer's Register,* VIII (April, 1840), 246-247.

[113] E. Emmons, *Report of the North Carolina Geological Survey,* 78-126.

[114] M. Whitney, *Soil and Civilization,* 11; E. Emmons, *The Swamp Lands of North Carolina,* 17. A contemporary remarked in 1859 that Liebig had contributed more towards putting agriculture on a scientific, "a sure and enlightened basis," than any other reformer. Daniel Moreau Barringer, *Address Delivered Before the Mecklenburg Agricultural Society,* 22-23.

[115] *Farmer's Journal,* I (April, 1852), 19. The editor advertised that he would analyze soils "in the most correct manner" for $5 and for an additional $5 he would furnish advice as to the proper application of manures to bring the

place too high hopes in soil analysis as a cure-all, and, when the cure was not immediately effected, began to think of the procedure as another "humbug."[116]

On the other hand, many farmers were led to give more and more attention to manuring in an effort to put back into the soil those elements now known to be essential to plant growth. In the search for "manures" to accomplish that purpose hardly anything was overlooked. For example, the *American Farm Book,* published in 1850, gave an appraisal as to chemical composition and function of each of the following when used as manures: ashes, lime, marl, shell sand, green sand marl, gypsum, bones, phosphate of lime, salt, nitrate of potash and of soda, charcoal, guano, soot, barnyard manures, poudrette (human feces or "night soil" as it was called), blood, flesh, hair, fish, cotton seed, peat, sea weed, rich turf, swamp muck, and clover, cow peas, and other green crops.[117] During the period 1830 to 1860 a major portion of the space in the agricultural press was taken up by articles, essays, and testimonials in which the fertilizing properties of the above substances were set forth with advice as to methods in their employment. With the "improving" farmer, the accumulation of manures became a year around activity.[118]

By experimentation, it was learned that the efficiency of various manures was increased by mixing them together into compost heaps. Professor J. J. Mapes, editor of the *Working Farmer,* was the foremost exponent in the United States of the composting procedures and practically every farm journal in the land carried one or more articles by him on the subject.[119] In Edgecombe County, which became the center of manuring activity in North Carolina after 1840, the compost heap became "the very best evidence of the improving man" and prodigious quantities of manure were applied to the soil. John S. Dancy, of that county and an ex-president of the state agricultural society, said in 1855 that he knew of several farmers in Edgecombe who had applied that year over fifty thousand loads of manure each to their lands. Each load

soil to its highest productivity. *Ibid.,* I (August, 1852), 159. The more scientific leaders, however, like Ebenezer Emmons, pointed out that a proper soil analysis could not be made so cheaply or simply. E. Emmons, *The Swamp Lands of North Carolina,* 20-22.

[116] E. Emmons, *The Swamp Lands of North Carolina,* 17-18, 22-23.

[117] R. L. Allen, *The American Farm Book,* 35-76.

[118] D. M. Barringer spoke as follows to the Mecklenburg County farmers: "*Manures* and *composts*—the true philosopher's stone to the farmer; make them, get them and save them. Get all you can; waste none; study the various characters of different soils and crops; accomodate [*sic*] and apply the manures accordingly." D. M. Barringer, *Address Delivered Before the Mecklenburg Agricultural Society,* 10. One of the strongest appeals for a generous use of manures to be found in the contemporary press was made by Paul C. Cameron in an address to the Orange County Agricultural Society in 1855. Quoted in the *Arator,* I (June, 1855), 73 ff.

[119] See reprint of article from the *Working Farmer,* by Professor Mapes in *Farmer's Journal,* I (April, 1852), 16-17.

contained about five bushels. As to the effects of manuring he said: "The improvement in the appearance of the soil, the increased product, and the high price of land in Edgecomb[e] tell their own tale."[120]

A small quantity of Peruvian guano was imported into Baltimore in the 1820's by John Skinner, editor of the *American Farmer,* and its properties became known, but it was not until after 1843 that the use of guano became widespread. The shipment for 1843 was quickly sold for seven cents a pound.[121] Imports of this fertilizer rose steadily after 1843 until they reached a pre-Civil War peak of 163,000 tons in 1854, from which point they declined to 1860.[122] Baltimore was the principal port of entry for guano, and, in 1854, the demand there exceeded the supply four times over.[123]

Guano attained popularity rapidly in North Carolina and its use became widespread.[124] In 1852, the price ranged from $40 to $70 per ton delivered in the state.[125] With only slight variations, these prices held for the remainder of the period. Because of its ease of application and almost miraculous effect when properly applied to crops, there was a tendency among some farmers to rely on guano to the neglect of other and more available manures, but which required far more labor in their accumulation and application. This practice aroused the criticism of agricultural reformers.[126] D. M. Barringer urged the farmers of Mecklenburg County to use homemade manures "which will answer as well, probably a great deal better. . . ."[127] Ebenezer Emmons said that fertilizers should be bought rather than attempt to cultivate the staples without them, but he strongly urged that the farmer avoid the expense by collecting and using the manures available on the farm.[128]

[120] *Arator,* I (August, 1855), 138. "Vito" expressed the opinion that nowhere in the world was as much labor being expended in the improvement of the soil as in Edgecombe County in 1855. "Most of our improving farmers," he said, "are raising from eight hundred to one thousand loads manure to the hand, and putting from one hundred to one hundred and twenty five loads to every acre cultivated in cotton." *Ibid.,* I (June, 1855), 79.

[121] A. O. Craven, *Soil Exhaustion,* 148-149.

[122] *DeBow's Review,* XXVII (September, 1859), 349. The decline after 1855 was caused chiefly by obstacles imposed by the Peruvian government rather than any lessening of the demand for guano. There was, however, some opposition to a too great reliance on it just as there was with regard to other articles imported into the South at that time. A contributor to *DeBow's Review* in 1858 spoke out against the "guano mania" and referred to guano as only a "temporary palliative." He also referred to a "Guano Convention" which was held in Washington that year to seek means of escaping what they called "the extortion practiced on the part of the Peruvians in reference to the sale of that bird dung." *Ibid.,* XXV (August, 1858), 147-148.

[123] *Ibid.,* XVII (July, 1854), 90-91.

[124] *North Carolina Farmer,* III (April, 1848), 252.

[125] *Farmer's Journal,* I (April, 1852), 13.

[126] For Solon Robinson's remarks on this point, see H. A. Kellar (ed.), *Solon Robinson,* II, 492-493.

[127] D. M. Barringer, *Address Delivered Before the Mecklenburg Agricultural Society,* 11.

[128] E. Emmons, *Report of the North Carolina Geological Survey,* 130.

A great deal of interest was shown in the patented manures in the latter part of the period. Chief among these was "Bommers," which used the compost method; and, according to claims made in its behalf, was "capable of converting expeditiously and economically, not only all kinds of vegetable substances, but also common earth, into rich and fertilizing manure. . . ."[129] De Burg's "Celebrated manure," which was made by a patented composting process similar to Bommers, also had some vogue in the fifties.[130]

The value of "green" manures in restoring soil fertility was recognized early and received increasing attention during the latter part of the period. Elisha Mitchell, in 1822, said that manures were valuable and should be used, but manuring alone could not be carried on extensively enough to restore fertility to the wasted soils of the state. He said:

> The great desideratum for the improvement of the Agriculture of North Carolina, is some vegetable of vigorous growth, which shall cover our fields from the burning sun during the heats of summer, and in the fall shall furnish a large body of vegetable matter to be ploughed in for the enrichment of the soil.[131]

Mitchell considered red clover unsuited for this purpose in North Carolina because of soil and climate factors which caused clover to ripen and decay too soon. He proposed that a standing prize be offered for the next twenty years for the best essay or discovery connected with the quest for this "vegetable."[132] Clover, however, was used for this purpose quite extensively by some North Carolina farmers. The Burgwyn brothers, of Northampton County, for example, sometimes planted as high as nine hundred acres of clover which they used both for pasturage and as a green manure.[133] Edmund Ruffin considered the pea better suited for "green manuring" in a southern climate than clover, "and nowhere does it grow better, with more certainty and more luxuriance than on the soils of Edgecombe." Yet, he found on a visit in that county in 1854 that, except as a secondary crop which was universally planted with corn, the manuring value of this "inestimable plant" was to

[129] Salisbury *Carolina Watchman,* August 23, 1849. In making this manure, leaves, straw, weeds, plants of various sorts, and dirt were mixed together to form the compost heap. At intervals, this mass was saturated with a lye solution to expedite fermentation. The resulting manure was thought by some farmers to be equal to Peruvian guano.

[130] Raleigh *Register,* April 15, 1857.

[131] Elisha Mitchell, "Agricultural Speculations," *Report on the Geology of North Carolina,* Part I, 54.

[132] *Ibid.,* Ebenezer Emmons also thought the climate of the state too hot and dry to offer much success in the culture of clover. E. Emmons, *Report of the North Carolina Geological Survey,* 181-184.

[133] Tarboro *Southerner,* June 25, 1853.

a great extent neglected.[134] Lucerne was also experimented with as a soil builder. W. J. Bingham, of Orange County, wrote in 1853 that he had been cultivating it for twenty years and considered it "of incomparable value in this climate both for soil feeding and for hay."[135] Ebenezer Emmons recommended lucerne as well adapted to the North Carolina climate, as a good forage crop, and said it "enriches the soil to boot."[136]

Closely associated with the program of the reformers for restoring fertility to the soil were proposals advocating proper drainage so that the fertility might be retained and the crop given the right kind of bed in which to grow. A contributor to the *Farmer's Journal*, in 1852, asserted that most lands in the state were improperly drained even though "it is . . . an established fact, that thorough drainage is the very first step towards improvement of the soil."[137] On the other hand, a visitor to the Edgecombe County Agricultural Society found that body discussing the subject as to whether land can be drained too dry for agricultural purposes.[138] With most farmers, drainage was effected by ditches, canals, and water furrows between seed beds. As in the case of fences and compost heaps, well-kept drains also became a sign of a "good" farmer. One planter in Edgecombe County installed a system of tile drainage in 1859.[139] Hillside ditching, contour plowing, deep plowing, and preparation of fields in beds were all concerned with better drainage as well as better tillage.

Rotation of crops, as a means of keeping the soil in a high state of productivity received a great deal of attention by the reformers.[140] George W. Jeffreys secured from a Virginia planter a rotation scheme which was widely used in Virginia, but there is little evidence that North Carolina farmers generally were interested in crop rotation.[141] There was considerable shifting of crops from field to field with each new planting season, but in most cases these shifts were not based on any scientific order. Edmund Ruffin, writing of a visit to Warren County in 1837, said: "There is nothing (here) worth the name of a *rotation of crops*";[142] but if the exhausting methods of tillage were not changed, said he, they would impoverish the land. On a trip through

[134] *Farmer's Journal*, III (October, 1854), 202; Edmund Ruffin, *Agricultural, Geological, and Descriptive Sketches of Lower North Carolina. . . .*, 89-91. A modern authority has declared the cowpea to be the most useful of the leguminous forage crops. Edward Burnett Voorhees, *Forage Crops for Soiling, Silage, Hay, and Pasture*, 253-264.

[135] *Southern Cultivator*, XI (April, 1853), 101.

[136] E. Emmons, *Report of the North Carolina Geological Survay*, 185.

[137] *Farmer's Journal*, I (July, 1852), 115-116.

[138] *Ibid.*, I (May, 1852), 50.

[139] *North Carolina Planter*, II (May, 1859), 147.

[140] For a contemporary discussion of the theory of rotations, see R. L. Allen, *American Farm Book*, 269-275.

[141] *American Farmer*, II (May, 1820), 61.

[142] *Farmer's Register*, V (August, 1837), 229-230.

Perquimans County in 1856, Ruffin found both the good and bad farmers neglecting rotation of crops.[143] "Alpha," of Edgecombe, said there was no system of rotation, "good or bad" used in that county in 1853; but some farmers did rest their fields in weeds, *"Just as much a crop* as any other," which if turned under at the proper time produced a "happy effect" on the soil.[144] A committee of the Edgecombe County Agricultural Society, in 1852, noted that no rotation of crops was being practiced in the county and recommended a six field system for both cotton and tobacco in which peas, clover, or weeds were suggested as fallow crops.[145] Regardless of the meager response, the farm press and agricultural leaders of the state continued to draw attention to the benefits to be derived from a scientific rotation of crops. The great increase in the culture of the grasses in the latter part of the period may be partially attributed to the widening interest in rotations.

Reclamation of worn out lands was another subject which received considerable attention from the reformers. In his "Agricultural Speculations" written in 1821, Elisha Mitchell said that most of the best lands in the state had already been exhausted and that farmers were being driven to cultivate "such inferior ridge-land as our Ancestors would have passed by as not worth cultivating."[146] Mitchell said the time had come, or was not far distant, when the farmer must decide whether it was less expensive to restore the old lands or to continue cultivating them with ever-decreasing crops. The real danger, as he saw it, lay in a too-close adherence to old methods which had wasted the land and would serve as obstacles to all plans for revival. He considered efforts to overcome that danger a proper objective of agricultural societies, and he strongly endorsed them.[147] More and more, the attention of the reading public was drawn by the agricultural press to the "glorious improvements" in restoration of worn out lands being made by English and Scotch farmers, and by the "shrewd and enterprising sons of New England."[148] As Solon Robinson wrote, the farmer was being invited to look "a little deeper beneath the surface for that quality of the soil

[143] Edmund Ruffin, *Agricultural, Geological, and Descriptive Sketches of Lower North Carolina,* 89. In a speech before the North Carolina Agricultural Society in 1856, Ruffin emphasized the weed and insect killing benefits derived from rotation of crops. *Carolina Cultivator,* II (November, 1856), 278. In this connection, too, see D. M. Barringer, *Address Delivered Before the Mecklenburg Agricultural Society,* 12.

[144] *Farmer's Journal,* II (September, 1853), 182. For other remarks on the suitability of various crops for use as fallow crops, see E. Emmons, *Report of the North Carolina Geological Survey,* 71-74.

[145] *Farmer's Journal,* I (June, 1852), 63-70.

[146] Reproduced in *Report on the Geology of North Carolina,* Part I, 51. It is interesting to note that in the cultivation of just such lands the most significant development in the agriculture of this period occurred in Caswell County, in 1839, when the Slade brothers first produced "bright" tobacco.

[147] Reproduced in *Report on the Geology of North Carolina,* Part I, 51-52.

[148] For example see *North Carolina Farmer,* I (June, 1845), 2-3.

that God must have placed there, when he said the earth should bring forth every fruit in its season."[149]

Michael Holt, of Orange County, wrote in 1832 that for twenty years he had been experimenting in an effort to improve exhausted lands, using his own ideas and those of others "worthy of notice."[150] Solon Robinson, in 1849, found Thomas P. Burgwyn "an enthusiast" on the subject; the latter kept his guest up past midnight "conversing exclusively upon the subject of improving and rendering fertile the worn-out lands of North Carolina and Virginia."[151] Burgwyn and his brother, Henry K. Burgwyn, did more than just talk and write about the subject. When they came into possession of their plantations in 1840 they found the lands exhausted and worn-out by the methods long used in their cultivation. By introducing the most up-to-date practices of deep plowing, drainage, heavy manuring, crop rotation, and use of soil improvement crops they developed their plantations not only into profitable enterprises, but also into show places.[152] By following similar methods, farmers put new life into their soils. "Vito," of Edgecombe County, increased his crop between 1848 and 1854 from five to fifty-five bags of cotton, and his corn from just enough to meet his own needs to two hundred barrels. In this period, he only increased his work force by the addition of one hand and an overseer. Except for a small quantity of gypsum, he used homemade manures entirely.[153]

Toward the latter part of the period, farm leaders began to place greater emphasis on diversification of crops. The principal purposes of this program were to provide a greater degree of self-sufficiency and to off-set, in so far as possible, the disadvantages North Carolina had as a cotton producing state.[154] As noted by a contemporary critic, the most neglected crops in North Carolina agriculture were grasses and meadows, small grain, garden vegetables, orchards and fruits, and stock animals. The attention of farmers was invited to the desirability of increasing these crops.[155]

Numerous articles appeared in the farm press on the subject of seed selection; they indicated a growing interest in the qualty of produce. Much advice was given as to how to select seed for planting. One editor advised: "Never grow a bad variety of anything, if you can help it. It takes the same room, and wants the same attention as a good

[149] H. A. Kellar (ed.), *Solon Robinson*, II, 492-493.
[150] J. G. deR. Hamilton (ed.), *Papers of Thomas Ruffin*, II, 59.
[151] H. A. Kellar (ed.), *Solon Robinson*, II, 224-225.
[152] *Ibid.*, II, 224-230, 492.
[153] The *Arator* quoted by the Tarboro *Southerner*, June 16, 1855.
[154] *American Farmer*, VIII (September, 1826), 194; D. M. Barringer, *Address Delivered Before the Mecklenburg Agricultural Society*, 16-18.
[155] *Farmer's Journal*, II (October, 1853), 211.

one."[156] To meet the demand for better seed that resulted from this program, some few North Carolina planters, notably Ebenezer Pettigrew and Josiah Collins of Tyrrell County, found it profitable to devote some of their effort to the production of new seed varieties. Each of these men developed new varieties of wheat and corn which commanded a premium price in the state for use as seed.

The tendency of farmers to plant more land than their labor force could properly cultivate or harvest, "over-cropping" as Charles Fisher called it, was common in the state and roundly condemned by the reformers. Fisher considered this one of the principal causes of the depressed and run-down condition of North Carolina agriculture in the decade of the twenties.[157] D. M. Barringer spoke of the "evil" of having too much land. His advice to the farmer was: "to make and keep his *land rich—tend it well, reject no improvement because it is new*"; and, if these things could not be accomplished on as much land as he wished to till, then he should "tend *less* and make it more productive for the labor bestowed."[158] The editor of the *Farmer's Journal* said "overcropping" was almost a universal error "even in the far famed county of Edgecombe," and he cited illustrations to show the practice to be less profitable.[159] An "improving" Edgecombe farmer, who, with thirty hands, had collected and distributed 22,000 loads of manure in one year on his 1,100 acres of cleared land, admitted that he owned too much land for his labor force. "I cannot tend more than one third of it," he wrote, "and that not tilled in the way I wish."[160] Another critic said that nine-tenths of the farmers cultivated too much land.[161] The edior of the *Arator* urged on his readers the paramount importance of having no more land in cultivation than could be well tended. "Be a FARMER," he wrote, "not a mere earth scraper, lazily scratching up sufficient earth to destroy the face of the soil, and throw seed away, or you will always have to scratch hard for a living."[162]

John Sherwood, editor of the *Farmer's Advocate,* was one of the leading advocates of greater efficiency in farm operations. "What you undertake, *do well,*" he urged.[163] Farmers were encouraged, in the interest of efficiency, to keep business-like records of their activities to include notations of stock on hand, implements, methods used, weather

[156] *North Carolina Planter,* I (January, 1858), 32. For examples of articles on the subject of seed selection, see Hillsboro *Recorder,* February 9, 1848, February 21, 1849; Salisbury *Carolina Watchman,* June 15, 1844; *Arator,* I (June, 1855), 81; *Farmer's Journal,* II (August, 1853), 140.

[157] *American Farmer,* III (August, 1821), 170.

[158] D. M. Barringer, *Address Delivered Before the Mecklenburg Agricultural Society,* 12-15.

[159] *Farmer's Journal,* I (July, 1852), 112-114.

[160] *Ibid.,* I (July, 1852), 102.

[161] *Ibid.,* III (June, 1854), 80-81.

[162] *Arator,* I (January, 1856), 301.

[163] *Farmer's Advocate,* I (March, 1839), 218-219.

conditions, time of planting, time and methods of culture, and of all experiments conducted.[164] This aim towards greater efficiency became the basis for many "Golden Rules of Good Farming," "Fixed Facts of Agriculture," and "Agricultural Maxims" which filled the farm journals.[165]

Agricultural leaders sought to instill in farmers a professional pride and to dispell the idea that "anyone can be a farmer," or that "farming is a dull, clod-hopping business." John Sherwood, in his *Farmer's Advocate,* was most forceful and persistent in advocating this feature of the reform program.[166]

As Avery O. Craven has said of the ante-bellum South, there was nothing peculiar about the problems reformers found in ante-bellum North Carolina agriculture. "Stated in simplest form," he wrote, "their task was one of introducing methods among those who tended the soil which would give profits without destroying fertility and which, in regions where already the earth's fullness had been spent would restore lands to fruitfullness."[167] Though not by any means to be found on every farm, or, perhaps, even half of them, there is abundant evidence that a substantial measure of reform had been accomplished in North Carolina agriculture by 1860. The will to improve had been aroused. Evidences of that may be seen in the increase in per acre yield of crops, increase in use of fertilizers and manures, improvements in farm implements, improved livestock, culture of new crops—particularly the grasses, production of premium fruits, decline in emigration, improvements in the farmsteading, greater crop diversification, increase in farm income, increase in land values, and better tillage of crops. The statistical story of this progress is reserved for a later chapter.

But because of the conservatism, superstition, and ignorance of many farmers, reform moved very slowly. As "New Hanover" noted at the time, it required "line upon line and precept upon precept" to induce many of them to give up old ideas and practices and adopt new ones.[168] Undoubtedly, a great number of farmers in North Carolina in 1860 were completely oblivious of the fact that a "spirit of reform" was in the air. "A Virginia Farmer" raised the question, which to him was unanswerable, as to why "bad" farmers, who had the example of the "good" before them, did not reform. ". . . but they don't do it," he said, "and

[164] *Farmer's Journal,* I (January, 1853), 305; *Southern Agriculturist,* II (March, 1842), 138-140.

[165] For example, see *Farmer's Journal,* I (August, 1852), 132, 158; II (August, 1853), 140; *North Carolina Farmer,* III (October, 1847), 97; *Carolina Cultivator,* I (March, 1855), 25.

[166] For one of Sherwood's glowing characterizations of the farmer, see *Farmer's Advocate,* I (August, 1838), 18.

[167] A. O. Craven, "The Agricultural Reformers of the Ante-Bellum South," *American Historical Review,* XXXIII (January, 1928), 302.

[168] *Farmer's Journal,* I (May, 1852), 34.

if we may judge the future by the past, they never will."[169] Perhaps, the chief criticism that may be levelled against the reform agencies is their failure to attract a wider participation in their programs by what may be called the great mass of "dirt" farmers. Yet, the meagerness of participation and contribution by that group did not give concern at the time. In accordance with the philosophy of the day, would not the benefits of the reform movement "trickle down" to them? Ebenezer Emmons urged "the utmost patience" for fear that discouragements might lead to the abandonment of "a right road" just because "the end is yet hid in mist and doubt."[170] To those reformers who expected a revolution in agriculture, the editor of the *Arator* sounded a note of caution. "While aiming at much," he wrote, "we must be encouraged with little, and labor for more with patience, diligence and perseverance—having our purposes strengthened with the sentiment, that

> 'Not to go back is somewhat to advance
> 'Men must learn to walk at least before they dance.' "[171]

[169] *Ibid.*, II (October, 1853), 216-217.
[170] *Legislative Documents* (1852), 13.
[171] *Arator*, I (April, 1855), 2.

CHAPTER VI

# COTTON AND TOBACCO, 1820-1860

As was noted in an earlier chapter, the culture of short staple cotton became widespread in North Carolina after 1800. By 1820, not only had the geographic limits in which the crop could be brought to successful fruition been determined, but farmers had turned to it in such large numbers that soon fears were being expressed that too much acreage and attention was being devoted to it.[1] The fact was that both large and small scale producers had found cotton to be the best money crop yet introduced. Several other factors played a part in this development. Cotton was found to be well adapted to cultivation by slaves. Little skill was required in its culture, and employment was provided for the entire slave family throughout the year. No expensive implements were required and large and small producers shared equal advantages in its culture. The product, too, was non-perishable, suffered little from transportation, possessed great value per unit of weight, and found a ready sale in the market places.[2] Cotton, therefore, as noted by the editor of one state paper, early "became a considerable object of culture. . . ."[3]

By trial and error, much information was quickly gained as to methods in the cultivation of cotton which was disseminated throughout the region in which it was grown. As was the case with most crops, however, uniformity of practice was never obtained.[4] Differences of

[1] In answer to an inquiry on the point, Calvin Jones, of Wake Forest, in 1828, questioned whether the culture of cotton had added much to the wealth and happiness of the country. "It occupies so nearly the whole year," he wrote, "as to stand much in the way of all agricultural improvements." Quoted in *American Farmer,* X (November, 1828), 265. In a committee report to the House of Commons in 1828, Charles Fisher, of Rowan County, lamented the fact that North Carolina was not producing all of her own foodstuffs. "The fact is," he wrote, "that the culture of cotton has misled us from our true interest and caused us to neglect every thing else." Quoted in *ibid.,* IX (January, 1828), 347.

[2] E. Q. Hawk, *Economic History of the South,* 235; U. B. Phillips, *Life and Labor in the Old South,* 128; U. B. Phillips, *American Negro Slavery,* 207-208; Matthew Brown Hammond, "The History of Cotton Planting in the South," *The South in the Building of the Nation,* V, 197-208.

[3] Wilmington *Gazette,* October 10, 1799.

[4] For a contrast of methods used in the culture of cotton in North Carolina about 1820, see article by "A Planter," of Johnston County in *American Farmer,* II (February, 1821), 373-374; George W. Jeffrey's "Memoir on the Culture of Cotton," reproduced in Denison Olmsted, *Report on the Geology of North Carolina,* Part I, 73-84; Robert Moore, "On Planting Cotton," reproduced in *ibid.,* Part II, 143-155; article by Calvin Jones reproduced in *American Farmer,* X (November, 1828), 265.

opinion arose between growers as to the best variety to plant, how to prepare the soil, how wide to make the rows, when to plant, how soon the plants should be thinned, how to cultivate, whether to top the plant or not, what fertilizers to use, and other aspects of its culture. Discussion of these differences continued throughout the ante-bellum period. With the slaveowning planters, too, a continual question confronted them as to how many acres should be planted per hand and still leave a sufficient labor force to raise the necessary provisions for the plantation.

Statistics of annual production are not available and the fact that most North Carolina cotton was marketed in Virginia and South Carolina makes it difficult to arrive at an estimate as to production prior to 1840. The trend of production, however, seems to have been upward until the late twenties, when, as the state's disadvantages as a cotton producer became more apparent, a sharp decline set in and very little cotton was grown in the state for the next fifteen years.[5] Remarking on this trend, the editor of the Raleigh *Register* said, in 1828, that there was a disposition "generally prevalent amongst the farmers in this state" to abandon, in some measure, the culture of cotton, "which is now unprofitable and likely to be more so," and turn to other crops such as grapes, sugar cane, olives, and silk worms.[6]

With reference to the causes of this trend away from cotton in the late 1820's, the editor of the *Southern Cultivator* said in 1860: "Thirty years ago, North Carolina abandoned the growth of Cotton, owing to causes not very clearly defined, but such as that the season was too short, the 'spring too backward,' soil too thin, & c. . . ."[7] Better reasons, perhaps, could have been offered, such as declining profits due to the fall in the price of cotton after 1819, continued heavy marketing costs, exhaustion of land due to poor farming methods, heavy emigration of population and capital which the state suffered between 1820 and 1840, low morale of the farmers, increase in the cost of living for those who over-specialized in cotton, mounting farm indebtedness growing out of the increased use of credit in anticipation of the cotton harvest, and the inability of the state to compete on even terms with the states of the Lower South in the production of the crop.[8]

[5] L. C. Gray, *History of Agriculture,* II, 677, 686, 889. For a contemporary estimate of production of cotton in North Carolina for the years 1834-1849, see *DeBow's Review,* XVII (October, 1854), 428-429.

[6] Quoted in *Niles' Weekly Register,* XXXV (September 13, 1828), 40. The North Carolina Board of Agriculture sought to promote the culture of these crops in the 1820's. Several articles were published on them and many grape vines were distributed by that body throughout the state. That this program had some success is indicated by the fact that North Carolina led the nation in the production of wine, according to the 1840 census.

[7] *Southern Cultivator,* XVIII (November, 1860), 350.

[8] In this connection, see articles by Charles Fisher in *American Farmer,* IX (January, 1828), 346-348, 353-354; and Calvin Jones in *ibid.,* X (November, 1828), 265-266, 273-274.

Consequently, because of the reasons indicated above and climatic limitations, cotton production in North Carolina failed to become the all-absorbing agricultural interest that it did in the more southerly states.[9] The number of planters who devoted their interest to its culture remained relatively small in comparison with the other "cotton" states. . Actually, the per cent of the American cotton crop grown in North Carolina steadily declined after 1800 and was smaller in 1859 than in any previous year.[10] This, however, does not mean that cotton production was unimportant to the economic life of the state or that the cotton product was negligible. Where the comparative per cent of the whole crop was declining in the state in the late ante-bellum period the number of bales being produced was increasing annually.

For the region in which it could be grown, cotton had superior advantages as a "money crop," but even in that region the number who placed their whole reliance on it were few. For example, John S. Dancy's plantation, "Panola," located in Edgecombe, the leading cotton county in the state in the fifties, produced in addition to 219⅓ bales of cotton in 1854, twenty-seven stacks of oats averaging 1,000 pounds each, 2,000 bushels of sweet potatoes, more than enough pork to meet the plantation's needs, and other subsistence crops.[11] Foolish, indeed, would have been the ante-bellum North Carolina planter to depend wholly upon outside sources for subsistence for either man or beast. That would have been taking too many chances and contrary to the conservative spirit of the people.

Beginning about 1840, evidences began to appear that a revolution was taking place in the methods of producing cotton and in agriculture generally in the state. This movement began in Edgecombe County and spread during the remainder of the period throughout most of the state. According to an eminently competent observer, this movement was chiefly characterized

[9] Cotton requires a long growing season of from six to seven months in order to mature all of its fruit. The dates of the last killing frosts in spring and the first in the fall are important in determining the geographic limits within which the crop can be profitably grown. A heavy frost stops the growth of the plant and prevents the maturing of the unripe fruit. Cotton also requires a warm or even hot atmosphere during the growing season. These factors, more than differences of soil, prevented the development of cotton culture in the Mountain and northerly sections of the state. Milton Whitney, "Climatology and Soils," *House Document No. 267*, 54th Congress (1896-1897), 144, 153.

[10] According to an estimate made to the House of Representatives in 1896, North Carolina's per cent of the American cotton crop in various years was as follows: 1799—11.2; 1809—8.5; 1819—6.2; 1829—4.7; 1839—6.5; 1849—2.9; 1859—2.7. *House Document No. 267* (1896-1897), 269.

[11] *Arator,* I (April, 1855), 14-15. This was doubtless the most famous cotton plantation in the state and many visitors, including Edmund Ruffin, were attracted to it. A reporter of the Warrenton *News* described it as the "most splendid plantation, take it all in all, that we ever examined." Printed in Tarboro *Southerner,* October 6, 1855.

By ditching and draining swamp land, naturally rich, but too wet to produce any crop; by using improved plows, and plowing the old fields up deeply; by *creating* manure for the poor barren sands; but principally by digging and spreading immense quantities of marl, or rather, seashell deposites [*sic*], which, until now, had lain idle and useless. . . .[12]

With variations in degree of emphasis to suit the particular circumstance, or crop, these practices formed the basis for a considerable reform in North Carolina agriculture during the remainder of the period.

To Edmund Ruffin, more than any other individual, credit was given for the improvements accomplished. Several farmers in Edgecombe were subscribers to his *Farmer's Register* and were familiar with his essay on calcareous manures. From this familiarity grew curiosity which was manifested by experimentation with the use of marl. Success followed.[13] Ruffin, who visited in Edgecombe in 1854, wrote of this development as follows:

Their success induced others to follow their example. Soon other materials were tried. One farmer began to make composts of earth and marl, and stable manures; another added ashes—a third cotton-seed—and others added other and smaller materials such as salt, gypsum and guano, but in few cases, and to limited extent.[14]

The collection of materials, composting, and distribution of manures, with the "improving" farmers, became the heaviest item of labor in the whole farming operation. "Latterly," wrote John S. Dancy in 1853, "composting every thing in the shape of animal and vegetable matter, received as much attention as the making of the crop."[15] Remarking on the changes that had occurred in the past ten years in this respect in Edgecombe, he said: "We can very safely assert that there was not as much manure *applied throughout the county then* as there is now *on several farms* that we are personally acquainted with."[16] In that year, Dancy himself kept two wagons in constant employment hauling up materials for the compost heaps, besides using his entire force for that purpose when the growing crops would permit and during the off seasons. In all, he collected, prepared, and distributed over 35,000 loads of manure (five bushels to the load) by broadcast over 350 acres of land. Most of this was put on lands intended for cotton. In addition, the cotton lands were also given an application of Peruvian guano. Dancy estimated that he would make from 900 to 1,000 pounds of seed-

[12] H. A. Kellar (ed.), *Solon Robinson,* II, 363.

[13] *Farmer's Journal,* II (April, 1853), 2-3.

[14] *Farmer's Journal,* III (October, 1854), 197. Of the system of compost manuring as carried on in Edgecombe, Ruffin said: "I am bound to admit, and fully believe, that the practice in general has been highly improving, beneficial, and profitable." *Ibid.,* 195.

[15] *Ibid.,* II (April, 1853), 2-3.

[16] *Ibid.,* 2.

cotton to the acre.[17] Actually, he made 1,245 pounds.[18] Edmund Ruffin, who saw this crop in growth, remarked: "The growth of the crop of Panola, was such as might be expected from a high degree of recent fertilization, and judicious culture. . . . All the crops were good, and of remarkable uniformity of growth. . . ."[19]

Heavy manuring alone did not accomplish those results. In Edgecombe, and the eastern counties particularly, proper drainage of the soil began to be recognized as a first step toward improvement.[20] Selection of better seed, use of improved implements, breaking the land deeply, and cultivating the crops thoroughly all helped increase the per acre yield and the appearance of farms on which these practices were used.

Remarkable results were obtained by those who adopted these practices in Edgecombe and elsewhere. Solon Robinson visited a cotton plantation near Tarboro which had had a steady increase in income of from $1,500 in 1844 to $4,200 in 1849.[21] And that was not all. "During the same time," he wrote, "complete new buildings and fences have been erected, and the value of the land more than doubled, and more than quadrupled in productiveness."[22] This farmer also made all of his provisions and nearly all of his clothing.[23]

Edmund Ruffin, in writing of the agricultural reform in Edgecombe, referred to R. R. Bridgers and J. L. Horn, two cotton planters of that county, who, by improvements increased their crops as follows:

| *Year* | *R. R. Bridgers* | *J. L. Horn* |
|---|---|---|
| 1847 | 19 Bales | 27 Bales |
| 1848 | 33 Bales | 43 Bales |
| 1849 | 53 Bales | 54 Bales |
| 1850 | 88 Bales | 83 Bales |
| 1851 | 136 Bales | 137 Bales |
| 1852 | 185 Bales | 165 Bales |
| 1853 | 170 Bales | 182 Bales |

"There are many others," he wrote, "who in the last 10 or 12 years have, by compost manures, doubled their crops—fewer have tripled theirs, and still fewer, including the above named, have increased theirs fourfold."[24]

[17] *Farmer's Journal,* III (July, 1854), 114-115.

[18] *Arator,* I (April, 1855), 14-15.

[19] *Farmer's Journal,* III (October, 1854), 194. Ruffin complimented Dancy as "a ripe scholar and a fine farmer." *Ibid.,* 211. "Panola" plantation was frequently referred to as a "model farm." For a description of "Panola" by Ebenezer Emmons, see *Executive Document No. 13* (1852), 59 ff.

[20] *Arator,* I (June, 1855), 79-80.

[21] H. A. Kellar (ed.), *Solon Robinson,* II, 363.

[22] *Ibid.*

[23] *Ibid.*

[24] *Farmer's Journal,* III (October, 1854), 199-200. In response to a request from Judge Thomas Ruffin, president of the North Carolina Agricultural Society,

In the county at large, production increased from 1,500 bales in 1840 to 3,097 in 1850; and, stimulated by higher prices in the fifties, reached 19,138 bales in 1859.[25] In 1850, Edgecombe was surpassed in production by three counties in the state, one of which, Anson, produced over three times as many bales. In 1859, Edgecombe's production almost doubled that of Halifax, her nearest rival.[26] A correspondent of the Goldsboro *Telegraph* wrote in 1852: "The true secret of farming seems in this county to have been found out, that is, when a crop is taken off to leave the land in better condition than it was before."[27] Commenting upon the progress made in Edgecombe, an editor said ". . . the great secret of this result is *thorough draining, deep plowing, and compost manuring*—means in the hands of every farmer in North Carolina."[28]

Although the progress made in Edgecombe received far more publicity, the revival of the cotton industry in the state was not by any means confined to that county. Newspapers, farm journals, and travelers gave wide publicity to the developments occurring there and soon "book farmers" all over the state were trying the new practices on their cotton lands. The result was that cotton production in the state experienced a phenomenal increase between 1840 and 1860—particularly in the latter decade. For example, only two counties produced more than 4,000 bales in 1850. In 1859, that number had increased to twelve. In Lenoir County the increase was from 185 bales in 1850 to 4,283 in 1859. At the same time production in Pitt increased from 171 to 7,634.[29] In commenting upon the progress made in that county, a Pitt farmer wrote in 1860: "Cotton seems to be the mania here, and all the larger planters have been, and are yet devoting the most of their time and attention to its cultivation. The crop this year will be much larger than ever before planted."[30]

Undoubtedly, in 1860, cotton was king of the money crops for North Carolina farmers. The reasons for this are fairly obvious. Possessing ideal staple qualities it sold at prices—particularly in the last decade of the period—which brought substantial profit to the producers. Al-

---

Bridgers prepared an account of his methods in making compost and an appraisal of the fertilizing value of various manures. Among other things, Bridgers used tremendous quantities of wood ashes. To secure these, trees were "belted" and burned on the new grounds just as in colonial times. Bridgers was a firm believer in the value of soil analysis as a means in determining chemical deficiencies. "We are not opposed to book farming here," he wrote, "we say read all the agricultural books that can be had; read to reap from the experience of others; read to understand the reasons of facts drawn from our own experience. . . ." J. G. deR. Hamilton (ed.), *Papers of Thomas Ruffin,* II, 480.

[25] *U. S. Census of 1860* (Agriculture), 105-111.

[26] *Ibid.*

[27] Reprinted in the Tarboro *Southerner,* March 20, 1852.

[28] *Arator,* I (June, 1855), 83.

[29] *U. S. Census of 1860* (Agriculture), 105-111.

[30] *North Carolina Planter,* III (August, 1860), 263.

though few planters could tell how much it cost to raise it, the best evidence indicates that it ranged near six cents a pound.[31] With the price averaging slightly over ten cents a pound during the decade of the fifties, cotton growers experienced a prosperity which some believed would be continuous. The cotton grower was further benefited by the fact that the cost of production was steadily declining.[32] Under such circumstances, there can be little wonder that the cotton producer sought out the best lands and the "likeliest" slaves. With him, cotton raising was his business and they were necessary factors in order to make the greatest success.

In this connection, an examination of the records of the Stephen A. Norfleet plantations, which were located in Bertie County and which have been previously referred to in this study, clearly reveals the tendencies to which the late ante-bellum North Carolina cotton plantation were subjected.[33] Norfleet operated three farms in Bertie—the "Home Farm," "Indian Woods Farm," from which he got all the receipts, and "Woodburn," from which he received two-thirds of the receipts and paid two-thirds of the expenses. During the decade of the fifties, although the plantations were not devoted exclusively to cotton, the acreage planted in that crop was increased annually.[34] From around a hundred acres in 1850 the increase was to 425 in 1860. This necessitated an increase in the landholdings. By eight separate purchases, Norfleet added more than 1,000 acres to his estate for which he paid out $14,197.23.[35] In the same period, by natural increase, inheritance, and purchase, the number of slaves was doubled.[36] In all,

[31] Solon Robinson made an exhaustive study in 1848 to determine the cost of production in South Carolina and Alabama. After adding marketing cost, which the producer had to pay, and interest upon a fair valuation of the property, he concluded that the cost of production in South Carolina was five cents and four mills per pound; and, in Alabama, was five cents and two mills per pound exclusive of freight and commission, but including interest on the investment. "My own opinion is," he said, "that whenever cotton is below six cents it does not pay interest on the capital invested, except perhaps in some few cases." H. A. Kellar (ed.), *Solon Robinson,* II, 241.

[32] L. C. Gray, *History of Agriculture,* I, 476. A contemporary claimed that in the space of half a century the cost of cotton had been reduced to about one tenth of its former value by the use of slave labor in its production. J. D. B. DeBow, *Industrial Resources,* II, 110.

[33] Stephen Andrews Norfleet Papers, Southern Historical Collection, Library of the University of North Carolina, cited hereinafter as Norfleet Papers.

[34] Norfleet received considerable income during the period from the sale of pork, hams, lard, corn, wheat, oats, fodder, and brandy. Norfleet Journal, *passim,* Norfleet Papers.

[35] *Ibid.* Five of the parcels totalled 1,010 acres. The acreage of the other three purchases was not indicated.

[36] Norfleet supplemented his labor force by the annual hire of Negroes, both free and slave, and by seasonal employment of others. The number of hired laborers was never large. They were used in digging ditches and clearing new grounds principally. Except for a blacksmith and occasional carpenters, these laborers were unskilled. He also paid his own slaves for such services as "extra labor in saving the crop," "extra work," "ditching," and "securing ashes for the crop." His list of black polls totalled sixty-four in 1858. *Ibid.*

$28,960 was paid out in the purchase of slaves and $1,250 received from the sales made. Each year after 1851 the plantations produced a handsome profit—amounting to almost $10,000 annually from 1858 to 1860.

As to the methods used, there was nothing in the practices Norfleet followed that would give him distinction over any other "improving" planter. He subscribed to agricultural papers and no doubt was familiar with the various techniques used in cotton production.[37] Like other planters, Norfleet's methods showed some slight variations from year to year and sometimes from field to field.[38] In its major aspects, however, the process of raising cotton suggested a routine which made the crop one particularly well adapted to culture by slave labor.

During the period to which reference is made, Norfleet kept a diary in which he made day by day notations of the manifold activities carried on on his plantations. By checking the entries for several years, which noted the appearance of the first bloom or the date the first bale was ginned, it becomes apparent that a time schedule for the many operations in cotton growing were followed with remarkable regularity. With the exception of time spent on other crops, or lost because of inclement weather, cotton raising for Norfleet and his slaves was a twelve-months-in-the-year business.

The following extracts concerning cotton raising, taken from Norfleet's diary for the year 1858, illustrate the operation of what is considered a typical ante-bellum North Carolina cotton plantation.

February 16: "Repairing fences & making compost of cotton seed & swamp muck hauled out previous fall. Grubbing & ditching."
February 25: "Grubbing and piling brush in newground."
March 13: "6 single plows barring off cotton ridges."
——— 15: "Man and 2 boys cutting down stalks in advance of plows."
April 20: "Sowing cotton: 9 plows bedding, 4 plows opening ridges and 4 plows covering."[39]

[37] Numerous articles appeared in the agricultural press setting forth the advantages of various methods in cotton production. Most of these possessed no real distinction. A few, however, were widely republished and, no doubt, influential with cotton growers. Chief among these was an essay by Colonel James M. Chambers, editor of *Soil of the South*, which was reprinted in J. A. Turner, *The Cotton Planter's Manual*, 11-20; and, an essay on the subject which was delivered by Whitemarsh B. Seabrook to the South Carolina Agricultural Society in 1843. Seabrook was president of that body. For a reprint of his essay, see *DeBow's Review*, I (April, 1846), 290-320.

[38] For example, Norfleet's diary contained no indication as to whether he topped his cotton or not in 1858. In 1859 and 1860, however, several remarks were entered showing that it was being done in those years. One authority, however, spoke of topping as "a contingent operation, and doubtful in its effects upon the crop." J. A. Turner, *Cotton Planters Manual*, 20.

[39] Colonel James M. Chambers, in his widely published essay on the cultivation of cotton, said cotton should be planted in drills rather than hills because it was more difficult to get a good stand in hills. Rows should be run off so as to give the plant as much sunlight through the day as possible. "The cotton is decidedly a sun plant," he said. The distance between rows depended upon the strength

——— 24: "155 acres planted in cotton & 120 yet to plant—which will be deferred for several days that the whole crop may not be ready to chop out at the same time."

May 1: "Cotton sown first beginning to come up."

——— 19: "Commenced siding cotton with 13 plows."

May 24: ". . . the wiregrass over-running the ridges in some places makes it necessary to work it out at once to save the stand of cotton."

——— 29: "Hoe hands chopping out cotton—worked over 220 acres in 5 days carrying 20 rows at a drift—or 2⅕ acres per day for each full hand."[40]

June 12: "Finished chopping out cotton—which completes the 1st working. Cotton Plows followed the Hoes."[41]

——— 14: "Cotton plows began to throw out middles. Began weeding cotton 2nd time."[42]

——— 28: "First cotton bloom."

July 12: "Completed 2nd working of cotton."

——— 31: "Cotton plows throwing out cotton middles. Hands weeding cotton for 3rd time."

August 2: "18 plows throwing out cotton middles last time."[43]

——— 6: "Exceedingly hot—crops literally burning up. Cotton is withered most of the day and the forms are dropping."

——— 10: "Dry & very dusty. The crops of both corn and cotton have suffered as much in the last week as I ever saw them in the same length of time."

——— 16: "On Friday last a heavy rainfall. The crops of every description have improved. . . . Commenced to haul mud from swamps today to make manure."

——— 23: "Unloading a boat with lime."

September 3: "5 men digging marl. 4 mauling fence rails."

---

of the soil. Rule: "When the crop is at maturity, the branches of the stalks ought slightly to interlock every way." Reprinted in J. A. Turner, *Cotton Planter's Manual,* 13-14.

[40] After the chopping was completed, Norfleet threw dirt to one side of the cotton ridge using a short mould-board. Several days later he followed with the same plow on the other side. Colonel Chambers referred to cotton as "this delicate plant." There were two schools of thought as to which should precede in the first working of the cotton—the hoe or the plow. Chambers recommended putting the hoe first in smooth lands and the plow in rough lands. All writers on the subject agreed that cotton should be kept free of weeds and grass. *Ibid.,* 11, 14-16.

[41] Cotton was planted uniformly in a furrow "by some careful hand" just thick enough to get a stand all along the row. *Ibid.,* 14-16.

[42] Usually the plows preceded the hoes in this working. The hoe hand had the "delicate duty" of further thinning the stand to the number of plants desired. Much of this thinning was done by hand. "The cut-worm and the louse are charged with many sins," wrote Colonel Chambers, "which ought to be put down to the account of careless working, at this critical stage of the crop." *Ibid.,* 16-17.

[43] The later workings of cotton were less close to the growing plant and shallower and intended to keep the soil soft and clean. There was some difference of opinion among planters as to how many times and how late cotton should be worked. *Ibid.,* 16-19.

——— 7: "5 horse carts hauling it (marl) on to the Old Pasture at the rate of 100 bushils per acre."

——— 16: "3 double plows breaking stubble land."

——— 18: "Peter whetting gin saws."

——— 23: "Hands picking cotton."[44]

——— 27: "Most hands picking cotton—saving the Petit Gulf cotton first for the seed, started the gin. The running gear is working tolerably well. Weighed 100 lbs, seed cotton—ginned it in 11 minutes—seed weighed 66 lbs.—leaving 34 lint cotton."[45]

October 2: "1 hand, Wilson, picked 236 lbs. which is the heaviest task I have ever had picked in one day . . . several picked over 150 lbs."

——— 5: "Joe Bragg picked 260 lbs, Wilson 256 lbs."

——— 6: "I think scarcely a third of bolls have yet been picked."

——— 11: "Cotton packing very badly because of dryness of the atmosphere. With difficulty the bales can be made to weigh 400 lbs."

——— 12: "Started picking over cotton crop second time. 54 bales ginned to date."

——— 14: "Very fine weather for saving cotton."

——— 18: "Most hands picking cotton."

——— 21: "80 bales ginned. Half the crop not yet picked."

——— 22: "Commenced to scatter piles of lime."

——— 26: "Most of hands picking cotton."

——— 27: "Seven men hands unloading a barge of oyster shells at my lower landing. 858 bushels in all."[46]

——— 29: "Most of hands picking cotton. Saving now about 500 lbs. cotton pr. acre."

November 2: "Most of hands picking cotton—averaged 130 lbs. each."

——— 3: "Picked yesterday 6,000 lbs. cotton—largest amount picked in one day. Joe Bragg picked 314 lbs."

——— 5: "114 bales ginned to date."

——— 6: "3 horse carts hauling ashes to highland."

——— 17: "Thermometer 29° after sunrise. Most hands picking cotton."

——— 25: "Cotton in new ground yielding a "disappointing" 700 lbs. per acre."

——— 26: "Some cotton being picked over for 3rd time."

——— 30: "Mauling rails—picking cotton—ditching—getting shingles. 186,500 lbs. cotton picked & 1675 bbls. corn."

[44] Picking was begun as soon as enough cotton was open to give profitable employment to the hands. In order to save the crop, every hand, old and young alike, was used in picking. To a considerable extent, the number of available pickers determined the acreage to be planted in cotton.

[45] As seed selection began to receive more attention by planters, the agricultural papers began carrying advertisements of numerous varieties of cotton for which outlandish claims were made and prices asked. Colonel Chambers recommended that each planter select from his own crop "the best formed and best bearing stalks." "We have been humbugged a great deal by dealers and speculators in this article, yet we would greatly err to conclude that no improvement could be made." J. A. Turner, *Cotton Planter's Manual,* 18-19. For the various seed varieties, see *ibid.,* 94-128.

[46] Norfleet prepared a kiln in which these shells were burned in order to make lime for use on his land.

December 4: "Double plows breaking ground for cotton."[47]
——— 8: "Carts hauling oyster shells."
——— 9: "Grubbing and scattering heaps of marl."
——— 10: "Picking cotton."
——— 13: "Hauling & distributing marl—75 bu. per acre."
——— 17: "Breaking land, ditching, picking cotton. . . ."
——— 20: "Making compost heap of cotton seed, marsh mud, lime and ashes."
——— 30: "Hauling marl. Breaking land."
——— 31: "Turning up lands for next years cotton crop with 2 horse plows to expose wire grass roots to frost. . . . Part of crop still ungathered, but everything considered we have no cause to complain. The crop of every kind has been a fair one. The cotton crop the largest that has ever been gathered from this plantation in any one season." (163 bales had been ginned up to December 23. The last of the 1858 crop was ginned on February 12, 1859 making the total 171 bales, or 245 pounds of lint cotton per acre. "A superior crop," noted Norfleet.)

Two overseers were employed during the year at a total cost of $400. Receipts from all crops amounted to $12,306.63, out of which expenses amounting to $2,015.59 were paid.

Although frequent reference was made in the agricultural press to the ravages of cotton pests and diseases, Norfleet made no mention of them in his records. Perhaps, like Colonel Chambers, he considered them "Providential Contingencies . . . about which little is known," and nothing could be done.[48] Whether that was the case or not, the ante-bellum cotton grower had his product reduced by losses to the caterpillar, cut-worms, lice, wilt, rust, rot, and boll-worms.[49] The boll weevil, whose depredations so severely curtail present day crops, did not invade the South until after 1860. Prior to 1860, no material progress was made in reducing the losses inflicted by pests and plant diseases.

To offset another source of loss to cotton growers, the development of processes for extracting oil from the seed began to attract attention. As has been noted, the excess seed was widely used in making composts and as cattle food, but this did not entirely solve the problem. By 1833, reports began to appear that machines for separating oil from the seed had been invented in Georgia and Virginia and uses for the oil were suggested.[50] The development of a suitable process for clarifying the

[47] As to the time and mode of breaking up land for cotton, Colonel Chambers wrote: "All lands for cotton, ought before the crop is planted, to be broken deep, close and soft, and this to be done enough before planting, to allow the rains gently to settle them." J. A. Turner, *Cotton Planter's Manual*, 13. Norfleet started breaking his land for the next cotton crop as early as November. When the weather permitted, this work was carried on throughout the winter months.

[48] J. A. Turner, *Cotton Planter's Manual*, 20.

[49] *American Farmer*, II (February, 1821), 373-374; Salisbury *Carolina Watchman*, September 25, 1846. See also U. B. Phillips, *American Negro Slavery*, 211.

[50] Reprint from the *Southern Banner* in Tarboro *Press*, August 30, 1833.

crude oil for illuminating and lubricating purposes failed, however, and the industry had no influence on ante-bellum cotton production.[51]

In concluding this section on cotton, emphasis should again be made to the facts that in North Carolina over seventy per cent of the farmers owned not a single slave, that among slave-owners the trend was towards smaller holdings; and, that the average size of farms was getting smaller as the ante-bellum period drew to a close. On the other hand, as was the case with Stephen A. Norfleet, there was a tendency to increase the size of the landholding among those who emphasized staple production. Since, in cotton raising, small and large-scale producers enjoyed about equal advantages none of the above exercised a detrimental influence on its culture. Unquestionably, a large proportion of the cotton produced in the state was by free white labor. Where it could be grown, cotton was increasing in popularity with producers in the state in 1860. Its culture was profitable.

Tobacco, the leading staple of colonial North Carolina, likewise experienced its ups and downs in the ante-bellum period. From 1783 to about 1840, several factors contributed to lessen the importance of tobacco as a staple crop and to reduce the acreage devoted to its culture. Chief among these were the loss of the monopoly of the British market, high import duties imposed by foreign countries, competition of foreign producers, the Napoleonic Wars, instability of prices, increase in cotton culture, soil exhaustion, and the influence of some agricultural leaders.[52] Tobacco, like cotton, sold in a world market. Therefore, many of the influences having a bearing on its culture were beyond local control.

When the colonies became free they gained the privilege of selling their tobacco wherever they chose, but lost the exclusive right of supplying the British market. Whatever advantage there was in that situation was short-lived; for, with the coming of war to Europe in 1792, those nations began imposing extremely high duties on all tobacco imported.[53] Once those taxes became a part of the national fiscal system, all effort to have them modified, or repealed proved unavailing. In fact, when changed at all, the tendency was to increase rather than decrease the rates. England, the best foreign customer down to 1860, in 1846 was collecting $72.75 on each hundred weight of unmanufactured tobacco imported.[54] In commenting upon this situation, a contemporary said:

[51] *North Carolina Planter*, I (November, 1858), 331.

[52] Meyer Jacobstein, *The Tobacco Industry in the United States*, 32-37, cited hereinafter as *The Tobacco Industry;* Meyer Jacobstein, "Tobacco Culture in the South," *The South in the Building of the Nation*, V, 163-168; A. O. Craven, *Soil Exhaustion, passim;* Joseph Clarke Robert, *The Tobacco Kingdom*, 135-136; L. C. Gray, *History of Agriculture*, II, 908-915; William Edward Uzzell, "History of Tobacco in North Carolina," 48, unpublished masters thesis, University of North Carolina (1929).

[53] M. Jacobstein, "Tobacco Culture in the South," *loc. cit.*, 163-164.

[54] *DeBow's Review*, II (July, 1846), 47. For a list of rates imposed by other countries see *Report of the Commissioner of Patents* (1843), 185-187; *Report of the Commissioner of Patents* (1859), 524-535.

"It will upon examination of this trade, be seen that Europe levies a revenue of about $30,000,000 on about 100,000 hhds. of tobacco, which cost in the United States about $7,000,000."[55] In view of the limitations such duties would impose upon demand, there can be little wonder that the amount of tobacco annually exported beween 1783 and 1840, except for a few unusual years, remained almost stationary.[56]

North Carolina producers felt the effects in a lessened demand for their tobacco and coöperated with other growers in efforts to better the situation. In 1840, the tobacco planters of Caswell County met in Yanceyville to select delegates to attend a national convention, to be held in Washington, to petition the United States government to intercede with other nations "to cause the duties now laid on our tobacco to be removed or in some degree taken off."[57]

The governor of Maryland presided over the Washington meeting. A report was prepared at the convention showing that "heavy and unequal duties are imposed on this article," a memorial was addressed to Congress, and an appeal was made to tobacco planters to request Congress for relief.[58]

In 1859, resolutions were introduced into the United States House of Representatives in which the foreign trade in tobacco was described as "clogged with restrictions and limitations, wholly inconsistent with that fair and reciprocal condition of commerce which ought to exist . . . and is, therefore, unsatisfactory. . . ."[59] These resolutions declared that it was the duty of the United States government to secure from foreign countries either a modification or removal of their import duties on tobacco.[60] These, and similar efforts failed, however, and, to 1860, tobacco continued to encounter this obstacle in finding markets broad.

Of greater significance to the American producers, however, were the limitations imposed on trade by England and France during the Napoleonic Wars and by the United States in the Embargo and Non-Intercourse Acts. These tended to stimulate the development of tobacco raising in other parts of the world—particularly Cuba, Colombia, and the East Indies.[61] This competition continued after the conditions out of which it sprang were removed.

In this connection, the period from 1790 to 1840 witnessed a tre-

[55] *DeBow's Review*, II (July, 1846), 53. A hogshead of tobacco weighed about 1,200 pounds.

[56] For a list of total annual exports of tobacco from the United States between 1790 and 1851, see J. D. B. DeBow, *Industrial Resources*, III, 348-349. Because of the Embargo, only 9,576 hogsheads of tobacco were exported in 1808. In 1814, because of the war, the number fell to 3,125. M. Jacobstein, "Tobacco Culture in the South," *loc. cit.*, 163-164.

[57] Raleigh *Register*, April 24, 1840.

[58] *Ibid.*, May 22, 1840.

[59] Raleigh *Register*, January 8, 1859.

[60] *Ibid.*

[61] M. Jacobstein, "Tobacco Culture in the South," *loc. cit.*, 163-164.

mendous expansion of area in the United States in which tobacco was grown. For example, Missouri which had not produced a single pound of tobacco to 1790, was estimated to have produced in the crop year 1844-1845 an amount which almost equaled the product of Virginia, Maryland, and North Carolina combined. Tennessee, Ohio, Kentucky, and Connecticut had also gone in for tobacco on a large scale.[62] This, of course, had tremendous consequences for the North Carolina producer.

As the competition of the western states developed the domestic price was forced downward. Ranging from ten to fifteen cents a pound between 1816 and 1819, the price fell quickly to eight cents when the panic of 1819 struck. From that point, the price declined almost continuously until it reached four and three-quarters cents in 1832. Then, it rose slowly to ten cents in 1839 and fell rapidly to four cents in 1842. From 1842 until 1850, except for special types of tobacco, the price hovered around an average of a little more than four cents a pound.[63]

As has been seen in the introductory chapter, tobacco required a virgin soil, which, under the system of culture employed, produced only a few crops before it lost its fertility and was relegated to less important crops or abandoned to grow up in briers and brambles. Land bore the major portion of the burden in that system of culture and the search for new lands continually widened the area in which tobacco was grown. When these lands became scarce, planters were faced with the same dilemma as the North Carolinian who wrote: ". . . our lands are nearly exhausted by continued cultivation . . . to us improvement is a matter of absolute necessity; we must fertilize our lands, or abandon our homes."[64] That many took the latter course, there is an abundance of evidence.

Because tobacco quickly wasted the soil, more than anything else, agricultural leaders, like George Washington and John Taylor, began to turn away from it. "It starves the earth," wrote Taylor, "by producing nothing to eat. Whatever plenty or splendor it may bestow on its owner, the soil it feeds on must necessarily become cadaverous, and its cultivators squalid."[65]

The principal reason for the decline in importance of tobacco growing to North Carolina, however, was the fact that after the invention of the cotton gin many planters found it to their advantage to give up tobacco and shift to cotton. As early as 1802, a traveler through the Piedmont section of the state noted that trend. "The low price to which tobacco is fallen in Europe, within these few years," he wrote,

[62] *DeBow's Review,* II (October, 1846), 252.

[63] For a list of average tobacco prices between 1802 and 1851, see J. D. B. DeBow, *Industrial Resources,* III, 349. See also, L. C. Gray, *History of Agriculture,* II, 914-915; J. C. Robert, *The Tobacco Kingdom,* 135-136.

[64] *Farmer's Register,* I (April, 1834), 642-643.

[65] From reprint of the *Arator* in *Farmer's Register,* VIII (December, 1840), 755.

"has made them give up the culture of it in this part of the country. That of green-seed cotton has resumed its place, to the great advantage of the inhabitants. . . ."[66] Of the two, cotton was more ideally suited to culture by slaves. It more nearly provided year around employment, required less skilled and careful management, and was less exhaustive of the soil.[67] Of major significance, too, cotton was increasing in profitableness to the producer, tobacco, decreasing.[68]

Notwithstanding the many factors that tended to reduce the importance of tobacco, its culture was never abandoned in North Carolina. The shift toward the West, which was noted in the colonial period, continued, and tobacco, rather than cotton, became the important money crop for the counties of Caswell, Granville, Person, Rockingham, and Warren. In the main, these counties lay beyond the fringe of the cotton-growing region and along the Virginia boundary. Their product was marketed principally in Petersburg with smaller quantities going to Farmville and Clarksville in Virginia.[69]

Methods employed in tobacco culture changed but little from those of the colonial period. Tobacco received less benefit from the labor-saving devices invented in the ante-bellum period than any other crop. In its culture, tobacco required more labor per acre, much of it painstaking hand labor, than any other crop. In this connection, John Taylor wrote:

> It would startle even an old planter, to see an exact account of the labor devoured by an acre of tobacco, and the preparation of the crop for market. . . . He would be astonished to discover how often he had passed over the land, and the tobacco, through his hands, in fallowing, hilling, cutting off hills, planting, replantings, toppings, suckerings, weedings, cuttings, picking up, removing out of the ground by hand, hanging, striking, stripping, stemming, and prizing, and that the same labor, devoted to almost any other employment, would have produced a better return than tobacco. . . .[70]

Omitted in the above, but very important from the standpoint of the tobacco crop and also as an employer of labor, was the preparation, planting, and care of the seed bed. Tobacco seed are even smaller than the proverbial mustard seed, and must, therefore, be first planted in specially prepared beds. This process called for the exercise of sound judgment and the expenditure of a great amount of labor. First, a site was selected which offered rich, friable, black virgin loam, or sandy

[66] R. G. Thwaites (ed.), *Travels West of the Alleganies Made . . . in 1802 by F. A. Michaux. . . .*, 298. The census of 1840 indicates that the eastern counties of Edgecombe, Hertford, Perquimans, Pasquotank, Nash, and Cumberland had practically abandoned the culture of tobacco.

[67] U. B. Phillips (ed.), *Plantation and Frontier Documents,* I, 92-93.

[68] M. Jacobstein, "Tobacco Culture in the South," *loc. cit.,* 164.

[69] L. C. Gray, *History of Agriculture,* II, 773.

[70] From reprint of the *Arator* in *Farmer's Register,* VIII (December, 1840), 755.

soil which was exposed to the sun and protected to the north. Trees were cut down around the bed site to permit the sun to enter. Then, great quantities of logs and brush were heaped on the bed and burned for the purpose of adding ashes and killing weed seeds and insects in the soil. This was done in the winter. The bed was prepared for the seed by repeated and thorough digging up by hand until the soil was loose and free of roots and clods. The seed were planted on the surface and rolled in during late winter. Protection against frost and the "fly," or flea beetle, was given by covering the bed with brush or, preferably, cloth. Towards the middle or last of May, the plants would be from four to six inches tall, at which time they were transplanted to fields prepared for them.[71]

The best tobacco was grown on new or fresh land. Except for the alluvial bottom lands, which were capable of producing crops for a longer period, the average tobacco land produced only three or four good crops before its productivity began to decline rapidly and the planter began preparing new grounds.[72] Tobacco did not respond to manuring as favorably as other crops. It was thought too much manure would be required, and, also, tobacco produced on heavily manured lands was coarser in fibre and less marketable.[73] It was more profitable to bring in new lands for tobacco. Consequently, very little manure was used in tobacco growing. "As to clover and plaster, lime and compost heaps," wrote a contemporary, "I hardly know a tobacco maker who uses them at all."[74]

In preparing new ground for tobacco, the first task was to remove the timber and brush. Usually this was accomplished by cutting, piling, and burning it on the land. This would leave the surface clean except for the larger stumps, which, in most cases, were ignored as much as possible. Then the land was closely plowed with a coulter, as deeply as two horses or oxen could pull, followed by hands with grubbing hoes who gathered up roots and debris for burning or removal from the land. This process was repeated one or more times, after which the land was

[71] Joseph Buckner Killebrew and Herbert Myrick, *Tobacco Leaf: Its Culture and Cure, Marketing and Manufacture,* 150-163, cited hereinafter as *Tobacco Leaf;* U. B. Phillips, *Life and Labor in the Old South,* 112. For a listing of varieties of tobacco grown in ante-bellum North Carolina, see J. B. Killebrew and H. Myrick, *Tobacco Leaf,* 34-45. For an account of transplanting, see *ibid.,* 169-178.

[72] From article on tobacco culture by "that excellent planter," Peter Miner, of Albemarle, Virginia, published in the *Arator,* I (February, 1856), 335-337. Methods differed slightly between planter and planter and between soils in tobacco as in other crops. For other "standard" articles on tobacco culture, see R. L. Allen, *American Farm Book,* 222 ff.; and article by Thomas W. Goodrich, of Greenville County, Virginia, reprinted from *The Southern Farmer* in *North Carolina Planter,* I (February, 1858), 61-62; Joseph Buckner Killebrew, *Report on the Culture and Curing of Tobacco in the United States,* 111-125, cited hereinafter as *Report on the Culture and Curing of Tobacco.*

[73] *Arator,* I (February, 1856), 335-336.

[74] *Southern Planter,* VIII (September, 1848), 259.

ready for hilling. These were prepared by hand with grubbing hoes—some effort being made to keep the hills aligned both ways and about three to three and one-half feet apart. In new land, the practice was to prepare the hills as early in the spring as possible in order to allow the soil to settle before planting.[75] In future plantings of this land, the original alignment of hills was kept and new ones were dug up on the sites of the old. The land in between was broken up by plow.

Depending upon the "season," the plants were set out between the middle and last of May. In preparation for this, the tops of the hills were cut off about four inches above their base. Replanting of hills where the plants had died was done after every rain until the last of June.[76]

Enough cultivation was given the growing plants to keep the earth light, and free of weeds and grass. The hoe was the principal implement used, but in lands free of stumps, cultivators and shovel-plows were also used. In the first working, a little earth was scraped away from the hills in order to kill the young grass. In future workings, earth was thrown by the plow and drawn up with the hoe around each hill. The number of such workings, as in cotton, depended upon the individual grower, but varied from two to four.[77]

The crop required considerable attention during the growing season. When the plants reached a proper size, "which observation and experience will readly point out," they were primed and topped. Priming consisted of stripping off four or five leaves at the bottom of the plant in order to leave about that many inches between the top of the hill and the first leaf. These leaves would otherwise be beaten down into the ground by rain, interfere with the cultivation, and, if matured, would have little market value. Topping was accomplished simply by pinching out the top of the plant—leaving the desired number of leaves. Some planters topped their tobacco two or three times—leaving eleven or twelve leaves the first time, nine or ten the second, and usually eight the last time. This prevented the development of seed and insured a more uniform ripening of the remaining leaves.[78]

Keeping the plant free of suckers, which sprang out from its base and right above the point where the leaves joined the stalk, was another process that had to be repeated several times. It was best to break these off when very young in order not to injure the plant. This, too, was done by hand.[79]

[75] *Arator,* I (February, 1856), 335.

[76] *Ibid.,* 336.

[77] *Arator,* I (February, 1856), 336.

[78] *Ibid.;* R. L. Allen, *American Farm Book,* 223; J. B. Killebrew, *Report on the Culture and Curing of Tobacco,* 115.

[79] R. L. Allen, *American Farm Book,* 223. When overgrown suckers were removed from the stalk they sometimes left wounds large enough to permit water to seep into the pithy part of the stalk causing decay. This disease was called

Of the various diseases and insects which attack the tobacco plant, the horn or tobacco worm is by far the most destructive. These worms are produced by a large, clumsy, gray colored fly, commonly seen around Jamestown weeds late in the evenings, which deposits its eggs at night on the tobacco leaf. These eggs hatch out in from twenty-four to thirty-six hours into small worms which begin to feed on the leaf immediately. Unless destroyed, these worms grow rapidly and devour the crop. Some planters maintained flocks of turkeys and guinea fowls to help keep down this menace in their tobacco fields. Most of the work of removing and destroying tobacco worms, however, was done by hand. This was slow work, but was repeated as often as necessary.[80]

The ground or cut worm also attacked the tobacco plant. This is the larvae of the common black bug usually found in large numbers under wheat shocks or similar covering. It was most destructive to the young tobacco plant. No practical solution was found to this menace.

Several diseases of the tobacco plant were known to the ante-bellum grower, but little was accomplished towards eradicating or controlling them. Perhaps, the most frequently mentioned was "spot" or "firing." Tobacco grown upon low lands suffered most from its attacks. It was thought to be an effect of a very moist season being followed by very hot weather. For that reason, complete and thorough drainage of tobacco fields was recommended. In this connection, a contemporary remarked: "It may not be amiss here to mention, that tobacco is more injured than any other crop by plowing or hoeing the ground when it is too wet, and to express a general caution on that head."[81]

In addition to the diseases and insects which menaced the plant, tobacco often suffered great damage from high winds, heavy rains, and hail. No guard could be set against these and when they struck, the planter could only hope for the best.[82]

When the plant reached its proper state of maturity, a point to be

---

"hollow stalk," caused the leaves to drop off, and was more apt to occur in crops planted in low grounds. *Arator* I (February, 1856), 336.

[80] *Arator,* I (February, 1856), 337; R. L. Allen, *American Farm Book,* 223. The tobacco worm has been described as "the great arch enemy of the tobacco plant," which sets a limit to the culture of the plant, and the control of which constitutes "the most disagreeable and tedious work attending tobacco growing." J. B. Killebrew and H. Myrick, *Tobacco Leaf,* 251. Flea beetles, wireworms, bud worms, crickets, and grasshoppers also fed on the tobacco plant. *Ibid.,* 243-258.

[81] *Arator,* I (February, 1856), 337. Plants growing in land not properly drained, also, sometimes took a diminutive growth, putting on many long, narrow leaves which were thickly set on the stalk. This was called "Walloon" tobacco and had no value. Of diseases of the tobacco plant, two recent scholars have said: "There is probably no crop produced of the same magnitude that suffers so little from disease as does tobacco, and nearly all these diseases may be avoided by proper care in the selection of the soils, in the judicious application of manure, and in the cultivation of the crop." J. B. Killebrew and H. Myrick, *Tobacco Leaf,* 238. For a listing of the various diseases to which tobacco is subject, see *ibid.,* 238-243.

[82] *Arator,* I (February, 1856), 337.

determined by the exercise of experienced judgment by the grower, but usually about three months after transplanting, the entire stalk was cut off and hauled to the barns for curing.[83]

Tobacco not only required more after-management than any other crop grown in the period, but it also, more than any other crop, reflected at all stages of its growth and processing the quality of management it had received. Growers who attained a good reputation for the manner in which they handled their crop always received the best prices in the market. Because of this, the acreage planted per hand was small in comparison with other crops.[84]

Because of its weight and size, tobacco was best cured by fire. Up to about 1812 most of the curing was done in the open air. This did not safeguard the leaf as well against sweating in the hogshead as was necessary and the process of curing in barns by wood fire was evolved. About 1837, charcoal was introduced as a fuel for this purpose.[85] The barns were usually built of logs, the cracks being closely chinked with mud so as to confine the heat.[86]

In these barns, the tobacco was hung after it was brought in from the fields. When the barn was filled, a log fire was kindled and closely attended for about forty-eight hours while the curing took place.[87] The tobacco was then taken down for stripping and stemming. In this process, the leaves were stripped from the stalk, assorted into three qualities, and tied into bundles of six leaves each. The third quality bundle—principally the ground leaves of the stalk—were stemmed before being bundled. The bundles were then whipped one against another to loosen and disengage any dirt that might be clinging to them, and the leaves straightened out. The bundles were then bulked in order to give them a set, that is, to make them "straight, close and nice." In doing this, each bundle was handled individually, being passed through the hand and squeezed enough to compress it into a small size. After remaining in bulk a short while, the bundles were hung up for prizing. This was the process of packing the bundles into hogsheads and was one requiring care and judgment. A contemporary, who described prizing as "the most difficult part of the business," wrote of it as follows:

> The order in which we strike down for prising is this—The stem must be so hard as to break freely two-thirds of its length beginning at the bigger end, while the leaf at the same time is so soft as to bear stretching open to its

[83] *Ibid.*, 336; J. B. Killebrew, *Report on the Culture and Curing of Tobacco,* 115-116.

[84] *American Farmer,* VII (April, 1825), 34. For information on the size of the tobacco plantation, size of the crop, yield per acre and per hand, see Appendices "B," "C," and "D" of J. C. Robert, *The Tobacco Kingdom,* 239-250.

[85] M. Jacobstein, "Tobacco Culture in the South," *loc. cit.,* 164-165.

[86] *American Farmer,* VII (April, 1825), 34; J. B. Killebrew, *Report on the Culture and Curing of Tobacco,* 116-117.

[87] For a description of the flues and curing by firing, see J. B. Killebrew, *Report on the Culture and Curing of Tobacco,* 116-117.

full breadth; this is the order which will enable it to go through a sweat and to make it show to the best advantage in the market, and to be best handled by the consumer or manufacturer.[88]

The bundles were put into the hogsheads with care, one at a time, with the heads pointing outward against the staves. In this manner the tobacco gave a neat appearance when uncased.

Beginning in the 1830's, evidences began to appear that the culture of tobacco in the state was on the upswing. With the exception of a few years, like that following the panic of 1837 which brought low prices, and the low price years of the early 1840's and again in 1849, the trend of production continued upward until 1859, when North Carolina produced the largest crop in its history.[89]

Several factors contributed to this increase in tobacco production. Improved flue-curing processes, in which charcoal was used as fuel and better temperature controls were employed, added to the quality of the product. The population was increasing rapidly with at least a proportionate increase in demand. New types of soil which produced peculiar qualities of leaf were discovered and exploited. Tobacco factories were first built and began giving unusually high prices for certain qualities of tobacco. Prices generally were higher—especially in the 1850's. Improvements in the transportation facilities encouraged growers.[90]

Under the stimuli of these influences, tobacco growing in the state took on new life. The area in which it was grown expanded and the crop increased tremendously—particularly in the 1850's. Counties such as Alamance, Forsyth, Franklin, Guilford, and Stokes, which had previously produced only small quantities, increased their production to such a point that they ranked among the twelve leading counties in the state in 1860. Older tobacco growing counties such as Caswell, Granville, Orange, Person, Halifax, Rockingham, and Warren doubled and trebled their product in the last decade of the period.[91] Benjamin Bailey, of Walnut Cove, noted "a strong disposition in many of the

[88] *American Farmer,* VII (April, 1825), 34.

[89] *Compendium of . . . the Sixth Census. . . . ,* 180-181; *The Seventh Census of the United States: 1850. . . . ,* 321; *U. S. Census of 1860* (Agriculture), 105-111.

[90] For example, under the caption, "Tobacco Good!", the Raleigh *Standard* of April 16, 1845, carried an account showing that tobacco was bringing from eight to fifteen cents a pound in the Petersburg market. A few months later, T. A. Yancey, of North Carolina, received thirty-five cents per pound for a No. 1. hogshead and seventeen and a half cents per pound for a No. 2 which were auctioned off in the Richmond market. *Niles' Weekly Register,* LXVIII (June 21, 1845), 256. See also article by "A Friend to Guilford" reprinted from the Greensboro *Patriot* in the Hillsboro *Recorder,* February 2, 1853; *DeBow's Review,* XIV (February, 1853), 175-176. For discussions of improvements in the curing processes, see J. C. Robert, *The Tobacco Kingdom,* 39-46; J. B. Killebrew, *Report on the Culture and Curing of Tobacco,* 116-117.

[91] *U. S. Census of 1860* (Agriculture), 105-111.

people to try it [tobacco]" in his section and requested that some "practical farmers" be induced to publish articles on its culture.[92]

The most significant development occuring, however, in the realm of tobacco culture, and doubtless, in the agricultural history of the state, was the beginning of production of "bright" tobacco by the Slade brothers, Abisha, Thomas, William, and Elias, in Caswell County in 1839. Paradoxically, this tobacco was grown on thin, sandy, ridgeland of a type previously thought to be practically worthless and unsuited to the culture of the crop. When cured, this tobacco had a beautiful lemon-yellow color, was light and dry, and particularly well adapted for use as wrappers and "fine cut" for pipe and cigarette smoking. The Slades used charcoal instead of wood in the curing process, and, at first, this was thought to have been the secret of their success.[93] Later, it was found that soil was the principal factor.[94]

Unusually high prices were received for this tobacco and its culture began to spread over the northern Piedmont section of the state and into Virginia. Captain A. Slade, of Caswell, and his two brothers, sold their entire crop of from 18,000 to 20,000 pounds, lugs included, to a Lynchburg manufacturer for the "extraordinary" price of thirty-five cents a pound. This was the product of some ten hands and meant an "unprecedented" income of about $700 per hand.[95]

The Slades readily communicated the experience they had gained in the culture of this tobacco.[96] The Civil War, however, cut short the trend of expansion in its growth. That had to await another day.

Although, in 1860, the state stood sixth in relative rank as a tobacco producer, the contribution of the state to the general development of the industry was extremely significant. Undoubtedly, the Civil War arrested a movement which would soon have brought the state into its later position of preëminence in the Union in the tobacco industry.

[92] *North Carolina Planter,* I (February, 1858), 57.

[93] J. B. Killebrew, *Report on the Culture and Curing of Tobacco,* 110-125; J. B. Killebrew and H. Myrick, *Tobacco Leaf,* II, 757; John K. Winkler, *Tobacco Tycoon: The Story of James Buchanan Duke,* 28.

[94] J. B. Killebrew, *Report on the Culture and Curing of Tobacco,* 110. Speaking of the soil on which bright tobacco was grown, two authorities said: "Poverty in the soil, for once, became the first principle of agriculture. The lands which grew the finest tobacco had light cream-colored soils, 93 per cent of which was siliceous matter. This porous, spongy, sandy earth, destitute of humus, and incapable of growing any crop without the most abundant application of manures, became the corner stone of a new agriculture." J. B. Killebrew and H. Myrick, *Tobacco Leaf,* 10-11. For an able discussion of the soil and climate requirements of bright tobacco, see Samuel Thomas Emory, *Bright Tobacco in the Agriculture, Industry and Foreign Trade of North Carolina,* 17-36.

[95] From reprint of article from the Clarksville *Tobacco Plant* in *Arator,* II (March, 1857), 718. Other North Carolina producers of bright tobacco were receiving prices almost as high in the Virginia markets. See quotations from those markets in Tarboro *Southerner,* August 15, 29, 1857.

[96] See reprint of article on the subject taken from the *Southern Planter* and appearing in *North Carolina Planter,* I (November, 1858), 331-332.

CHAPTER VII

# THE CEREALS, 1820-1860

Because of its highly acceptable quality as a food for both man and beast and its hardiness as a plant, corn was undoubtedly the most important crop grown in ante-bellum North Carolina. Unlike cotton, tobacco, and rice, which were confined to certain areas by limitations of soil and climate, corn could be and was grown in every inhabited part of the state. For example, eighty-four of the eighty-six counties into which the state was divided in 1859 produced, in that year, over 99,000 bushels of corn each, and only nine of the counties had a product of less than 200,000 bushels![1]

Certainly, corn was beyond challenge of any other crop in its claim for the attention of producers or in the immediate benefits bestowed upon them. It made up an important part of the diet of the people, both white and black, in the form of hominy, hoecakes, grits, corn pone, and mush. For the Negro slave, it was the most important single item of diet.[2] Likewise, draught animals, horned cattle, poultry, and particularly swine, consumed a large part of the corn crop. A few farmers cut immature corn for their livestock; most, however, used the matured grain, tops, fodder, and shucks in feeding. No other crop served such varied purposes; besides, corn whiskey was an important item of consumption and trade. There can be little wonder that over one-half the acreage devoted to farming in the South was planted in corn.[3]

Although most of the corn produced in the state was consumed at home, much of it found its way to markets elsewhere in the form of whiskey, corn meal, and fattened pork. Many planters, in the Roanoke River valley particularly, devoted themselves almost exclusively to its culture and sold their product by the ship-load in Norfolk, Baltimore, Philadelphia, New York City, and Providence.[4]

[1] *U. S. Census of 1860* (Agriculture), 105-111. The "barrel" was often used in measuring corn. It contained five bushels shelled or ten in the ear. *American Farmer,* X (November, 1828), 265. For the amounts of corn and other cereals produced in North Carolina in the census years 1840, 1850, and 1860, see the table at the end of this chapter.

[2] A traveler through the Piedmont section in 1802 remarked that corn was grown on every plantation and that the major portion went to feed the slaves—each slave receiving an allowance of two pounds per day. R. G. Thwaites (ed.), *Travels West of the Alleganies Made . . . in 1802 by F. A. Michaux. . . ,* 304.

[3] Thomas F. Hunt, "The History of Cereal Farming in the South," *The South in the Building of the Nation,* V, 214-215.

[4] Stephen A. Norfleet and Lewis Thompson of Bertie County and Thomas Pollock Burgwyn and his brother Henry King Burgwyn of Northampton County

As to methods in the cultivation of the crop, this period witnessed a considerable shift from the practice of planting in hills to that of planting in furrows or drills. Also, towards the end of the period, plows and cultivators were used nearly everywhere in its culture, and more attention was given to seed selection. The hoe, which had been the principal implement used in the cultivation of the crop, became secondary to the plow and cultivator.

Opinions varied as to the best methods to employ in every stage of corn growing, from the preparation of the ground to the harvesting of the crop. A committee of the Edgecombe County Agricultural Society, which had been appointed to look into this question in 1852, reported that it was ". . . rare to find in a neighborhood any two farmers agreeing, either in their manner of cultivating the corn crop, or in the particular variety of corn that they cultivate."[5] Methods were found to vary with soils, situations, and producers, and both good and bad crops were produced under each system. Doubtless, "the skillful and judicious farmer" was one, who, wedded to no particular system, varied his methods to suit his soil or soils.

The frequency with which the problem of methods in corn raising became the subject of articles in the agricultural press and the topic for discussion before agricultural societies not only reflects the great importance of the crop, but also a greater divergence of practice than could be found in the culture of any other crop. The adaptability of corn to almost every kind of soil being used for farming purposes, and the fact that corn could be relied upon to reward even the most careless and slovenly practices with a harvest, accounts for this situation. A complete failure of the corn crop hardly ever occurred except from drought, hail, and flood. When such catastrophes struck the corn crop, as they sometimes did, great distress and suffering resulted.[6]

The points of view expressed in the discussion held by the Scotland

produced great quantities of corn. Josiah Collins and Ebenezer Pettigrew of Tyrrell County raised corn almost exclusively as a money crop in the latter part [illegible]iod. Note will be taken of their successes later in the chapter.

[5] *Farmer's Journal,* I (August, 1852), 139-140.

[6] Stirling Ruffin, of Rockingham County, wrote Thomas Ruffin on August 12, 1819: "The crops here are ruin'd! In Stokes and the adjoining Countys in Virginia they are worse; corn cannot now be engaged at any price; and except for the plentiful crop of wheat, we should certainly be threatened with starvation next year." J. G. deR. Hamilton (ed.), *Papers of Thomas Ruffin,* I, 228. In 1826, a prolonged drought practically ruined the crops in the eastern part of the state. Corn went up to $1.50 a bushel. Many people abandoned their lands and set out for the West. The governor was petitioned to impose an embargo on the shipments of provisions out of the state as long as the scarcity continued. *Niles' Weekly Register,* XXX (August 19, 1826), 436-437; *ibid.,* (August 26, 1826), 449; *ibid.,* XXXI (September 2, 1826), 4. Floods often damaged the corn crops. The Roanoke River, for example, over-flowed its banks eight different times in 1857. This was the best corn-growing section of the state. Norfleet Diary, Norfleet Papers.

Neck Agricultural Society, on June 5, 1852, on the subject: "The cultivation of corn, embracing the method of preparing the land, and whether it is best to plant in single, double, treble, or quadruple beds, etc.," is an excellent illustration of the divergence of views held by farmers on just one phase of the culture of corn.[7]

Devereux said he plowed as much of his corn lands in winter as possible, breaking it up into treble beds, fifteen feet wide. These beds should never be reversed in succeeding years, but should be kept permanently the same. This practice had the effect of gradually deepening the water furrow as the bed became higher. Devereux bedded all kinds of land and found that narrower beds suffered more in periods of drought.

W. B. Smith had tried beds of from one to six rows on his swamp land and expressed the opinion that those twenty or thirty feet wide would keep as dry as single row beds. He, too, did not believe that the beds should be reversed. He supplemented the water furrow by preparing an open drain away from sinks or low places in his fields. This drain was always kept open.

Brinkley broke his land in single beds by throwing the soil to the water furrow of the previous year. This had the effect of keeping the land almost level. In larger beds, the rows near the center always looked best, and, for this reason, the single-row bed was favored. He remarked, too, that land broken into single beds would keep drier in wet weather and would not bake as badly in dry or windy weather.

Bryan planted his corn in double beds.

James Smith remarked that single beds suffered more in drought.

W. K. Smith explained that the middle rows in wide beds looked better because they were planted higher, but observed that the rows nearest the water furrows produced the most corn.

President R. H. Smith did not use beds on his upland corn, but planted it on the level. For the lowlands, he preferred the wide beds because of fewer water furrows to keep open, and because they furnished a warmer bed for the growth of the corn.[8]

In this connection, Josiah Collins, of Tyrrell County and one of the leading corn producers in the state, broke up his fields into single-row beds of from five to six feet in width. Collins' land was all low, having been reclaimed from the swamp, and was thoroughly drained by a network of canals and ditches for several years before planting a crop. In his system, the water furrow of one year became the corn row in the next planting. In so far as possible the land was bedded up in the fall for the

[7] At each meeting of this society, four members, who had been appointed by the president, held a "conversation" on a subject which had been selected at the previous meeting. The meetings were informal and after these members had finished their discussion, others were encouraged to "converse." *Farmers' Journal,* II (October, 1853), 212-213.

[8] *Farmer's Journal,* II (October, 1853), 212-213.

spring planting. Because of the rank growth of weeds, which in this rich, deep soil reached seven and eight feet in height, a preliminary operation was necessary before the ground could be broken. This consisted of sending the hoe-hands with heavy, broad hoes through the field cutting the weeds and corn stalks and laying them out length-wise in the water furrows. Then, two furrows were run with a turn-plow on each side of the water furrow throwing the slices so as to overlap atop the weeds and roughage. In this manner, the weeds were made to serve as a manure for the succeeding corn crop. Several weeks later the plows returned and continued throwing soil toward the weeds until the old beds had been reversed and new water furrows established on the line of the corn row of the past season.[9]

The principal development in the manner of planting corn was the widespread shift from the method of planting in hills to that of planting in drills or rows. Several factors encouraged this change. The increase in the use of animal drawn implements after 1800 brought a larger acreage into use. One man with plows and cultivators, planting his corn in rows, could tend a larger crop than a man using a grubbing hoe and planting in hills. Also, more of the planting was being done in old fields, which were freer of stumps and roots, and, therefore, easier to till with a plow. The agricultural reformers recommended the new system. "The most important point to be attended to in the cultivation of Indian corn," wrote George W. Jeffreys, "and upon which the success of all other operations depends, is deep ploughing." To the question as to how this was to be done, he answered: ". . . by cultivating corn in ridges and furrows."[10] Like most innovations in agriculture, this one encountered the opposition of "old fashioned" farmers. Sidney Weller, of Halifax County, which was one of the leading corn-producing counties in the state, was warned by a neighbor against using the new system as late as the 1840's. Weller, however, planted in drills and harvested what he thought to be one-fourth again more corn than he could have if planted in hills.[11]

Some corn growers, instead of planting in hills or in drills, planted on the level, in checks, in such a manner that in the later cultivation the

[9] *Farmer's Register*, VII (November, 1839), 727-729. Collins' largest crop up to 1839 was 8,000 barrels or 40,000 bushels. His next door neighbor, Ebenezer Pettigrew, also a leading corn producer and planter, broke his corn lands in the same manner. *Ibid.*, 730.

[10] *American Farmer*, III (April, 1821), 39.

[11] *North Carolina Farmer*, II (April, 1847), 253-254. Sidney Weller was, without a doubt, the most articulate planter resident in ante-bellum North Carolina. There was scarcely a farm paper published in the period which did not at some time carry one of his articles. His principal subjects were silk and wine production in both of which he claimed to have discovered an "American System" which would practically guarantee success. Weller was also an enthusiastic advocate of general agricultural improvements. See article by the author, "Sidney Weller: Ante-Bellum Promoter of Agricultural Reform," *N.C.H.R.*, XXXI (January, 1954), 1-17.

plows could operate at right angles. This system reduced the work of the hoe-hands to a minimum. In preparing the land for this system of planting, rows were run off across the field at a distance of from four to four and one-half feet from each other. Then, other rows were run perpendicular to these and at the same distance apart. Two or three grains of corn were then dropped at the point where the rows intersected.

This method had obvious disadvantages. Chiefly, no regard was had for the "lay" of the land or provision made for drainage. For these reasons, this system did not prove popular with the better farmers. "Mine," wrote Calvin Jones, of Wake Forest, "is to break and bed up new land as well as old; to plough on horizontal lines, as instructed by Randolph, and to plant corn and everything else in drills as prescribed by Taylor."[12]

The actual planting of corn took place generally in April and May. This was done by hand—about twice as much being dropped as would give a good stand—and a few inches of soil were drawn over it by the foot of the dropper.[13] Some farmers recommended dropping the corn in single grains rather than in twos and threes because it was easier to to chop it out to a stand with a hoe than to pull it out by hand.[14]

Several varieties of corn were grown in the state from which selection might be made. A correspondent from Albright wrote: ". . . there are several kinds raised among us, the goard seed, the yellow, the flinty ditto and the white; all kinds have their advocates, and no one kind appears to stand greatly pre-eminent among us."[15] More and more farmers were realizing, however, the value of proper seed selection, and, towards the end of the period, advertisements were appearing in the farm press setting forth the peculiar qualities of numerous varieties of corn.[16]

The size and shape of the ear, the number of grains to the ear, and the number of ears to the stalk were guides variously used in seed selection. This system was described as follows:

It is an invariable rule with me when I am housing my crop of corn in the fall, to select the largest, finest, and best filled ears which I can find. A person that will pursue this plan, may always keep his stock pure, and it

[12] *American Farmer,* X (November, 1828), 265.

[13] *Ibid.* George W. Jeffreys, of Person County, made what he called a "machine" for covering the corn. ". . . it is of quadrangular shape, light and easily managed by a boy—the fore part consisting of teeth, knocks off all of the clods that may be found on the ridge, whilst the hinder part covers the corn with the loosened earth, and all done at one stroke. . . ." *American Farmer,* III (April, 1821), 40.

[14] *North Carolina Planter,* I (April, 1858), 119.

[15] *American Farmer,* II (December, 1820), 311.

[16] "Baden" corn was, perhaps, the most widely advertised variety of seed corn. A Buncombe County farmer claimed to have produced over fifty bushels of this corn to the acre; whereas, he could harvest no more than thirty bushels to the acre of "our country corn" even though it was planted in his "best fields." *Carolina Planter,* I (February, 1840), 27.

will never degenerate. In seeding my crop, I never use more than half of the ear. The smaller end is invariably laid aside; the larger end is generally used, and it is always well filled.[17]

"J. G. A.," of Anson County, increased the productivity of his corn by selecting his seed only from those stalks which produced two or more ears.[18] This planter's technique in developing better corn foreshadows the production of the present day hybrid variety. He wrote:

> Another part of my experiment that I believe to be new, was this: During the time my Corn was tasselling and silking, and before the ferina had arrived at that state at which the fructification takes place, I went several times through my patch of Corn, and carefully examined each stalk, to see if the shoots on it promised to perfect more than one ear, and from every one that did not, the tassel (the male bloom,) was carefully drawn, in order that the variety which produces two or more ears, might not be fructified by the ferina, from that variety which produces only one ear. . . . it is indispensably necessary that the experiment be made remote from any other Corn, for certain it is, that the ferina may be conveyed by insects and the atmosphere to a considerable distance and thereby get mixed.[19]

Several planters evolved varieties of corn which they sold for seed and to which they usually gave their names.[20]

While only a few planters, such as the Burgwyns of Northampton, George W. Jeffreys of Person, and Ebenezer Pettigrew and Josiah Collins of Tyrrell counties made any attempt at a scientific rotation of crops including corn, many planters began manuring their corn lands and the practice grew in popularity. Various sorts of manure were used. The use of "green" manures by Collins and Pettigrew has been noted. A Burke County farmer thought hog manure best.[21] An Anson County grower used "crude" cotton seed, hog, and sheep manure on his corn, but considered sheep manure the best.[22] N. W. Woodfin, of Buncombe County, who won honors in the "Corn Sweepstakes" sponsored by the North Carolina Agricultural Society in 1853 for the best yield of corn per acre from reclaimed land, used on his prize-winning acre four hundred loads of muck and fifty loads of barnyard manure. This was scattered broadcast on the land after it had been first broken up. Then the land was turned under again to a depth of sixteen inches,

[17] *American Farmer,* X (January, 1829), 361.
[18] *Southern Agriculturist,* III (January, 1830), 14-15.
[19] *Ibid.*
[20] Ebenezer Pettigrew and Josiah Collins of Tyrrell County, for example, both developed varieties of good reputation. B. H. Wall, "Ebenezer Pettigrew," 273. For further comment on this subject see R. L. Allen, *American Farm Book,* 148-154; article by Calvin Jones in *American Farmer* reprinted in *Southern Agriculturist,* I (September, 1828), 423-424; *Report of the Commissioner of Patents* (1849), 142-143.
[21] *American Farmer,* III (May, 1821), 53.
[22] *Southern Agriculturist,* III (January, 1830), 14-15.

after which, fifty more loads of muck, fifteen of stable manure, and four bushels of plaster were mixed and broadcast on the land. This was harrowed in before planting. After the corn came up the land was given a top dressing of plaster and ashes.[23] This, of course, was an unusual amount. The average corn field received far less when it was manured at all. It may be said, however, that toward the end of the period more manure was being used on corn than ever before.

Many farmers had different ideas as to how the corn crop should be cultivated. As one expressed it: ". . . there is no uniformity of practice, for we have no system. . . ."[24] Just about every farmer had his own system. The main points of difference concerned the implements to be used and how many times and how late the crop was to be worked. Fundamentally, the main features of the various schemes were the same. About ten days or two weeks after the corn came up most planters ran a furrow on both sides of the row using a coulter, or a plow without a mouldboard.[25] On the other hand, some farmers, including Thomas W. Harris, of Halifax County, who had the reputation locally of making "more corn with less work than any planter," used the turn plow in the first working and threw the dirt away from the corn, leaving it on a ridge where it could be easily hoed.[26] The hoe-hands followed the plows in either case, chopped out the young grass, partially thinned the corn, and uncovered any that required it. The farmer who used the coulter usually followed with a cultivator, about three or four weeks later. This implement, drawn by one horse, stirred and loosened the earth, without exposing it to the sun, and killed weeds and grass growing in the middles.[27] The hoe hands followed, chopping out the grass, pulling away the suckers, and thinning the corn down to the desired stand. The mouldboard plow was used in the second working, also, in those fields where it had been first used. This time, a single furrow was run on each side of the row—throwing dirt to the corn. The cultivator followed, breaking out the middles.[28] In the later workings, regardless of the implements used, care was exercised lest the ground be broken too deeply near the growing corn. The crop was laid by when it became large enough to shade the ground and not be injured by any grass that appeared. The last working usually consisted of a single furrow down the middles with a double mouldboard plow or sweep.

[23] Disinterested parties measured a yield of 149 bushels and three quarts from this acre—the ears averaging about seven and one-half inches in length. *Farmer's Journal*, II (March, 1854), 372-373.

[24] *American Farmer*, X (November, 1828), 265.

[25] Sidney Weller wrote of this plowing: "The coulter I consider a *sina qua non* or indispensable as to drilled corn. It loosens deep and well for a fine start. . . ." *North Carolina Farmer*, II (April, 1847), 253-254.

[26] *North Carolina Planter*, I (April, 1858), 119.

[27] *American Farmer*, III (April, 1821), 40.

[28] *North Carolina Planter*, I (April, 1858), 119

This killed the grass and opened the water furrow which helped prevent washing.[29]

The practice of planting peas along the corn row was common throughout the state. Some planters alternated a row of corn with a row of peas in their fields.[30] Other crops intertilled with corn were, beans, watermelons, cantaloupes, and pumpkins.

To secure roughage for livestock, some farmers, when the corn had reached maturity but while the leaves were still green, cut the tops above the ears and pulled the leaves from below. This, along with the corn, was usually left in the field until it cured, after which, it was hauled to the barn for housing or stacked in shocks in the field.[31] On other farms, the entire stalk of matured corn was allowed to remain standing, to be harvested at the leisure of the grower, after which, the livestock were permitted to browse over it. Quite often, too, the entire stalk was cut and shocked in the fields for further curing and possible later housing in barns or sheds. Matured corn, perhaps more than any other crop, could endure the weather with a minimum of loss. Harvesting, therefore, could be more leisurely done.

From the few prize-winning acres that produced around 150 bushels, the yield of corn ranged downward to the point where scarcely an equal return was received for the seed planted. The average per acre yield in Orange County in 1850 was between fifteen and twenty bushels. "Some do much better. Others worse."[32] In Halifax County, one of the leading corn producing counties, the average yield per acre of the best farms in 1849 was about thirty bushels. This was thought to have been increased by drill husbandry and thicker planting.[33] In 1852, the average yield of uplands in Cabarrus County was estimated at around thirty bushels and of lowlands about forty.[34] This estimate was probably considerably exaggerated. The Burgwyn brothers, of Northampton County, who were farming some of the best corn land in the state and using the most advanced implements and techniques, produced an average of only thirty-one bushels per acre in 1848. This was an average yield for them.[35] Josiah Collins and Ebenezer Pettigrew, whose planta-

[29] *American Farmer,* III (April, 1821), 40. Some farmers believed that late working of corn often proved more injurious than helpful. *Ibid.; Farmer's Journal,* II (October, 1853), 212-213.

[30] *North Carolina Farmer,* II (April, 1847), 253-254; *American Farmer,* X (November, 1828), 265.

[31] Cutting tops and pulling fodder in this fashion has been declared a harmful practice which materially reduces the yield according to investigations of agricultural experiment stations. Thomas F. Hunt, "History of Cereal Farming in the South," *The South in the Building of the Nation,* V, 217-218.

[32] From reply of W. J. Bingham, of Orange County, to a circular from the office of the Commissioner of Patents. Bingham said the gourd-seed and Collins varieties were most commonly planted. *Report of the Commissioner of Patents* (1850), 376.

[33] *Ibid.,* (1849), 143.

[34] *Ibid.,* (1852), 93-94.

[35] H. A. Kellar (ed.), *Solon Robinson,* II, 228-229.

tions were located on reclaimed swamp lands in Tyrrell County, sometimes produced around sixty bushels to the acre on their rich, deep soils. The normal yield for them, however, was more often between thirty-five and fifty.[36] Frederick Law Olmsted, who was perhaps the most inquisitive visitor to the state in this period, made five random measurements in three different fields near Raleigh in 1853 and found the corn rows to be five feet apart, and the stalks in a row to be, on an average, "not over an inch in diameter at the butt," and standing two feet and one inch apart.[37] It might have been remarked that the soil in that section was not adapted to the production of even average corn crops.

Depending upon location and the state of the crops, the price of corn varied considerably from time to time.[38] Since corn was bulky and hard to transport, most planters sought merely to raise an amount sufficient for their own needs. Tremendous quantities were consumed in the state. Ebenezer Pettigrew, for example, used seven hundred barrels on his own plantation.[39] Throughout the state, corn was the most popular crop and, for most people, the "staff of life."

Stimulated by the heavy European demand for food stuffs after 1790, wheat growing became considerably more widespread and important in the economic life of the state. This was particularly true in the Piedmont section, which soon became the leading wheat producing part of the state—a distinction which it has not yet relinquished.[40] Although demand lessened considerably with the coming of peace to Europe, the passage of the Corn Laws in England, and other restrictions on trade; culture of the crop became permanently fixed in the agriculture of the state.[41] Just as in the case of corn, the census of 1860 disclosed that some wheat was produced in every county in the state.[42] Heaviest production was centered in the Piedmont section.

[36] *Farmer's Register,* VII (December, 1839), 728-730. Both of these planters harvested their corn into flat-boats, which were permitted by a network of canals to operate near each field. The corn was poled in to large barns, located on the shores of Lake Phelps, and lifted from the boats by water power into barns for housing. Pettigrew invented a water driven corn-sheller which shelled out sixty barrels an hour. *Ibid.,* 729-731; B. H. Wall, "Ebenezer Pettigrew," 402.

[37] F. L. Olmsted, *Journey in the Seaboard Slave States,* 320.

[38] L. C. Gray, *History of Agriculture,* II, 812-813.

[39] B. H. Wall, "Ebenezer Pettigrew," 278.

[40] L. C. Gray, *History of Agriculture,* II, 609.

[41] The panic of 1819 also contributed to the deepening of the depression of the wheat growers. A traveler through Orange County in that year wrote: "Wheat crops uncommonly good, price 25 cs. per bus. & little or no demand for it at that or even any price." J. S. Bassett (ed.), "General Slade's Journal of a Trip to Tennessee," *Historical Papers Published by the Trinity College Historical Society,* Series VI (1906), 39. The tariff duties imposed by European countries on wheat and other food crops between 1815 and 1840 ranged between 300 and 400 per cent. L. B. Schmidt and E. D. Ross (eds.), *Readings in the Economic History of American Agriculture,* 218.

[42] *U. S. Census of 1860* (Agriculture), 105-111. The best wheat soils, as a matter of fact, correspond very closely with the best corn soils. H. H. Bennett,

Randolph, Chatham, Davidson, Guilford, Rowan, and Forsyth counties led the state in the order listed. Interestingly, the number of slaves owned and the average size of slaveholding in those counties was not particularly high. Unlike cotton, tobacco, rice, and corn, once the wheat crop was planted it required no further working until the harvest. Therefore, even on the larger plantations, wheat was always grown in conjunction with other crops.[43] With Thomas Ruffin, wheat growers could say: "The variety of our productions is such that if one staff of life fail, nature will supply another in due season. . . ."[44]

Numerous varieties of wheat were grown in the state. Most popular of these were the Mediterranean, blue-stem, bearded, red spring, May, golden chaff, Black Sea, early purple straw, fly-proof, Troy, Iturean, Genesee, and Zimmerman's.[45]

In the main, the wheat crop was carelessly planted and the yield per acre was very low. Referring to the methods used in Orange County in 1850, W. J. Bingham wrote: "It is generally sown on corn land, and put in in a slovenly manner. Some do better, sowing on fallow ground broken up late in summer, or early in autumn, and ploughed or harrowed in."[46] R. H. Bennet reported that in Iredell County: "One bushel of seed is sown per acre; and usually dispatched, whether on stalk or fallow land, with one ploughing. Wheat, owing to the Hessian-fly, late frost, and, I may add, to our indifferent farming, is here an uncertain crop."[47] In Lincoln County, in 1851, it was reported that wheat raising was becoming more profitable, "but is mostly too carelessly put in."[48] There, the grain was sown broadcast on the corn lands, after the corn had been gathered in October, and lightly plowed in.[49] Joshua Harris, of Cabarrus County, wrote: "Our people are so

*The Soils and Agriculture of the Southern States,* 20. This was amply demonstrated in the successes with both crops had by such ante-bellum planters as Ebenezer Pettigrew and Josiah Collins of Tyrrell, and James Pollock Burgwyn and Henry King Burgwyn of Northampton counties. *Farmer's Register,* VII (December, 1839), 728-730; H. A. Kellar (ed.), *Solon Robinson,* II, 228-229.

[43] Thomas Pollock Burgwyn, of Northampton, who specialized in wheat and was one of the leading producers in the state, had on one plantation in 1849, in addition to 450 acres in wheat, 350 acres of corn, 520 of clover, 100 head of cattle, and hogs "unnumbered." The previous year he had produced 24,000 pounds of pork which had been fattened chiefly upon "wild potatoes," peas, pumpkins, clover, and soft corn. "The crops upon each of the other plantations, are upon nearly the same scale." H. A. Kellar (ed.), *Solon Robinson,* II, 229.

[44] J. G. deR. Hamilton (ed.), *Papers of Thomas Ruffin,* II, 169.

[45] *Report of the Commissioner of Patents* (1850), 368-369, 375; *ibid.* (1852), 93-94. Ebenezer Pettigrew, of Tyrrell County, experimented thirty-five years with various wheats and finally produced a variety which was well adapted to the soil and climate of the Coastal region. This variety was widely known and sold in that section as, "Pettigrew Wheat." B. H. Wall, "Ebenezer Pettigrew," 294-295.

[46] *Report of the Commissioner of Patents* (1850), 375-376.

[47] *Ibid.,* 368-369.

[48] *Ibid.* (1851), 315.

[49] *Ibid.*

engaged with raising cotton, that there is no improvement making in the wheat crops. . . . the good land is all planted in cotton. . . ."[50]

"The average yield is from 4 to 10 bushels to the acre," wrote a correspondent from Lincoln County, "and it is a good season in which we get that much."[51] In Cabarrus County it was reported as "about ten bushels."[52] And, of Orange County, a correspondent said:

> I think the yield per acre has been diminishing for many years, and hope, after falling to an average of 5 bushels per acre, it has reached its minimum, and is beginning to rise, by a gradually improved system of farming. A few of us have brought up old exhausted broomstraw fields to yield 8 to 10 bushels per acre."[53]

Of his harvest in 1855, Thomas Ruffin, of Alamance County, wrote:

"The yeild [*sic*] is . . . good making, I think, from 7 to 20 bushs. to the acre through the crop and probably averaging 14 or 15 bushs. It is the best crop I ever made. . . ."[54]

As may be seen in the illustrations cited above, the average farmer, in raising wheat, imposed upon his land the major portion of the burden of production just as was done in colonial times. Scarcely any farmer placed chief reliance on his wheat crop. Little, if any, manure was used under it. With the expenditure of a minimum of labor, the crop was planted usually in October, November, or December and given no further attention until the harvest, which took place in the first weeks of June. Grown in this manner, wheat's role was a minor one in the agricultural economy of the state, and whatever of harvest was produced was considered as clear profit.[55] Generally, the best lands and most of the manures and labor were reserved for the better paying crops of cotton and tobacco.

On the average farm, the cradle had about replaced the sickle by 1800 as the principal implement used in harvesting small grain crops. Some continued to use the sickle, however, throughout the period. Although well-to-do planters could afford the reapers and threshers, which came in after 1830, there is very little evidence that their use became widespread.[56]

[50] *Report of the Commissioner of Patents* (1851), 315-316.
[51] *Ibid.*, 315.
[52] *Ibid.* (1852), 93-94.
[53] *Ibid.* (1850), 375-376.
[54] J. G. deR. Hamilton (ed.), *Papers of Thomas Ruffin,* II, 472.
[55] The adjustment of a "good" farmer to this mode of raising wheat is well illustrated by the German who, in 1821, censured the farmers of all racial strains in the state for considering seven and one-half bushels of wheat per acre a satisfactory yield. Eleven years later, after experiences of his own in the business, he estimated that "a well-managed field of 30 acres should average that amount." William H. Gehrke, "The Ante-Bellum Agriculture of the Germans in North Carolina," *Agricultural History,* IX (July, 1935), 146.
[56] T. F. Hunt, "The History of Cereal Farming in the South," *The South in the Building of the Nation,* V, 219-220.

After cutting, which usually took place in early June, the wheat was tied into bundles, stacked in shocks in the field, and allowed to remain there for further ripening of the grain. Harvest time was a busy time on every farm.[57] Sometimes the wheat crop was seriously damaged after harvest, but before it had been threshed and properly housed. Threshing was accomplished by trampling out the grain under the feet of cattle, by flailing, and by use of the threshing machine. The threshing machine, operated by a portable horse-power device, which received its power from one or more horses on a tread-mill, became more common after about 1830.[58]

The wheat crop, like all others, was often reduced and sometimes totally destroyed by the depredations of pests and diseases.[59] Chief among these enemies of the wheat grower were the Hessian fly so-called because of its supposed introduction into this country by Hessian troops in the time of the Revolution, the weevil, joint worm, rust, smut, cockle, and cheat. Some made no effort to prevent the appearance of these destroyers in their crops.[60] On the other hand, the more intelligent growers sought diligently for protection. Recipes for the prevention or control of these diseases and pests were evolved and widely publicized in the agricultural books, journals, newspapers, and almanacs. Smut, perhaps, gave more trouble than any of the others and most of the recipes for its prevention directed that certain treatments of the seed be made before planting. "I secure it from smut, the great enemy to early wheat," wrote an Orange County grower, "by soaking it in a solution of salt, from 12 to 48 hours, and drying in lime or ashes, so as to scatter. This I have found a perfect protection."[61] Crop damage suffered from the ravages of insects and diseases posed an universal problem. Being looked upon by many as providential contingencies, it is doubtful that very much control was effected or damage prevented.

There were in the state a few wheat planters who rank with the largest and most progressive in the nation. Chief among these were

[57] J. G. deR. Hamilton (ed.), *The Papers of Thomas Ruffin*, II, 39.

[58] *Ibid.*

[59] *Farmer's Journal*, II (February, 1854), 344; *North Carolina Planter*, I (July, 1858), 207-208, 218; *ibid.*, (August, 1858), 244-245.

[60] *American Farmer*, X (November, 1828), 265. A recent student of the subject referred to plant pathology as an "infant among sciences," and said: "Prior to the year 1870 . . . there was practically no home-developed knowledge on these subjects (smuts, rusts, blights, mildews and rots), and to the great mass of farmers . . . plant diseases were mysteries. . . ." Beverly Thomas Galloway, "Plant Pathology; A Review of the Development of Science in the United States," *Agricultural History*, II (April, 1928), 49-51. The science of entomology was said by a correspondent of the *Genesee Farmer* to be "less understood than almost any other branch of rural knowledge. This general neglect of an important science is doubtless one cause of the alarming increase of destructive insects in many parts of the country." Reprinted in *Farmer's Journal*, II (March, 1854), 353.

[61] *Report of the Commissioner of Patents* (1850), 376. For other examples, see *North Carolina Farmer*, III (October, 1847), 113; *Farmer's Journal*, I (October, 1852), 202; *North Carolina Planter*, I (August, 1858), 244-245.

the Burgwyn brothers, Thomas Pollock and Henry King, of Northampton County. Doctor William R. Holt's operations in Davidson County were equally as progressive but not on so large a scale . All three attracted considerable attention by the success of their operations, and frequent reference was made to them in the agricultural press. Neither gave his whole attention to wheat raising and each won distinction in other phases of his agricultural activities, as has been previously noted.

The accomplishments of the Burgwyns' are particularly worthy of note because their lands had been previously "worn out" and it was first necessary to restore them. This was done by thorough drainage, extensive use of lime and other manures, resting the land in clover, part of which was turned under as a green manure, and by deep plowing.[62] Although started about 1840, their system was considered "new" by their neighbors, overseers, and even their slaves, all of whom predicted failure.[63] They did not fail, however, and by 1855 their plantations "Alveston," "Thornburg," and "Occonachee Wigwam" had become show places.[64]

The contrast between the Burgwyns' pains-taking and expensive method of seeding their wheat and that previously described in this chapter is striking. Henry King Burgwyn wrote his overseer, Arthur Souter, in 1843, as follows:

> As to the management of the . . . wheat, I have been on some wheat plantations on James River and have changed my plans in some measure. I shall use all my oyster shells on that field. . . .
>
> I wish you to follow this plan *exactly* in preparing that field for wheat: 1st Haul all the stable manure, pond mud &c on the field as soon as you *possibly* can (I think you might haul the stable manure as soon as you can cut down some of the corn so it will ripen & stack it so as to let the carts go thro & drop their loads.) As soon as you have got all this done and the corn gathered . . . turn in your manure, mud, leaves &c &c *headlands* & *all* except the road from the mill to the stable, in the gullies you must put down one or two rails & a little brush so as to make blind ditches & cover them over with earth deep enough to plough over without disturbing the rails . . . (you must try your best on this field for I expect great things from it next year) [.] As soon as the land is *turned over* the osyter shells must be hauled on it & burned in convenient piles so as be distributed and as soon as the ashes are cool enough they must be spread *on top* of the *ploughed* ground (& they will

[62] *Farmer's Journal,* I (August, 1852), 147; Salisbury *Carolina Watchman,* June 21, 1855; H. A. Kellar (ed.), *Solon Robinson,* II, 228-229.

[63] H. A. Kellar (ed.), *Solon Robinson,* II, 228-229; reprint from the Fayettevillle *Observer* in *Arator,* I (July, 1855), 121.

[64] Raleigh *Register,* July 4, 1855. A story was circulated that Henry King Burgwyn grew wheat so thick that "40 black snakes tried to go through one of his wheat fields, but were unable to do so. . . , and crawled over *'on the heads of the grain.'* " After reviewing the scope of operations of these plantations, Professor Phillips wrote, ". . . the Burgwyns . . . were a challenge to the wheat growers of any region in their day." U. B. Phillips, *Life and Labor in the Old South,* 127, 253-255.

sink in soon enough) at the rate of 30 to 40 bushels to the acre, or more if there is enough. I think you can do this by the 15th Oct. then in 3 or 4 days the shell ashes will be cool enough to sow your wheat on top and harrow in. & I shall sow not less than ½ bushels to the acre. . . .[65]

In this manner, the wasted fields were restored. By 1855, it was estimated that their 1565 acres, which were planted in wheat, would average producing about twenty bushels to the acre. Although, not an exceptional yield when measured by present day standards, it was undoubtedly considered so by those ante-bellum growers who believed a "well-managed" field should produce from seven to ten bushels.

The Burgwyns used the very best and most up-to-date implements and equipment in their farming. By 1855, they were using ten reapers in their harvest, and, besides, had five in reserve for use in case of break-downs. They tried several different makes but preferred the Hussey reaper to all others. These were drawn by four horses.[66] Their wheat was threshed by steam power in a mill capable of handling 1,000 bushels per day.[67]

Corn and wheat were the principal items produced for sale on the Burgwyn plantations. A visitor wrote of Thomas Pollock Burgwyn's sales in 1848: "His crop sold . . . from the labor of fifty hands, (besides ditching, manuring, and other improvements, and making all supplies of bread and meat, and part of the clothing for the people.) was 10,000 bushels of corn, at 45 cents, and 3,000 bushels of wheat, at 90 cents."[68] As the price of wheat rose during the 1850's more emphasis was placed on that crop. Henry King Burgwyn's wheat crop, of 1855, was reported at 900 acres on which he produced 23,000 bushels. This he sold in Petersburg for $2.25 per bushel.[69]

As the period drew to a close greater emphasis was being given to the raising of wheat. More growers were using the new implements and abandoning the careless practices of the earlier period. Benefits of these changes became apparent almost immediately. During the last decade of the period, wheat production in the state increased by over 100 per cent. Of the degree of comparative progress with other states, note will be taken later.

North Carolina's geographic location coincided with the northern limits of the belt in which rice could be successfully grown. Unlike wheat and corn, its culture was almost wholly confined to the lowlands

[65] Henry King Burgwyn to Arthur Souter, July 24, 1843, Burgwyn Papers.

[66] Raleigh *Register*, July 4, 1855; H. A. Kellar (ed.), *Solon Robinson*, II, 228-229.

[67] Raleigh *Register*, July 4, 1855; H. A. Kellar (ed.), *Solon Robinson*, II, 228-229. The Burgwyns' methods of harvesting and threshing their wheat were described as being "far ahead of anything else in North Carolina." *American Farmer*, VI (March, 1851), 307.

[68] H. A. Kellar (ed.), *Solon Robinson*, II, 228-229.

[69] U. B. Phillips, *Life and Labor in the Old South*, 253-255.

of the Coastal region. Along the creeks and rivers from the point where they began to rise and fall with the tide to the point where the water became brackish or salty, rice growing became an important, if uncertain, agricultural activity in colonial times. Although a hardy crop, rice growing was so confined and handicapped by climatic limitations, by the necessity for irrigating the crop, and by the unhealthiness of the locale in which it was grown that, once these facts were realized, it proved a poor competitor to cotton and other crops. The story of rice growing in ante-bellum North Carolina, therefore, is one that marks a gradual recession of the area in which the crop was grown, and the development of a great degree of specialization in its culture.

By 1800, short-staple cotton was already making heavy inroads on lands previously planted in rice. Cotton at that time was commanding a very high price, and it was thought that one good crop was equal to two good rice crops.[70] Ground was also lost to corn and wheat. For example, the Collins and Pettigrew estates in Tyrrell County, which were developed primarily as rice plantations, had by 1820, almost abandoned the crop for corn and wheat.[71] The same tendency existed throughout the rice-growing section of the state with the result that, at the close of the ante-bellum period, Brunswick County, alone, was producing over six-sevenths of the rice grown in the state.[72]

As a result of the specialization carried on, Brunswick County actually produced more rice in 1859 than the entire state produced in 1849 and more than twice as much as was grown in the state in 1839.[73] Having extensive marshes or tidal swamps with fresh water, and a more southerly location, Brunswick County made rice growing a specialty. Very little corn or other crops were grown—not even enough to meet the local demand.[74]

Columbus, Duplin, Sampson, and New Hanover counties followed Brunswick in that order in rice production in 1859. Columbus was a poor second, producing only about one bushel for each sixty-seven produced by Brunswick. New Hanover, which had a crop of approximately one and one-half million pounds in each of the census years 1839 and 1849, in 1859 produced only 69,000 pounds.[75]

In the nation at large, rice production reached its peak in 1850 and

[70] R. G. Thwaites (ed.), *Travels West of the Alleganies Made . . . in 1802 by F. A. Michaux*, 303. Perhaps the best illustration of the inability of rice to compete with the spread of cotton culture is to be found in the advertisement, of 1825, which offered the famous rice plantation "Orton," on the Cape Fear, for sale principally as cotton growing land. U. B. Phillips (ed.), *Plantation and Frontier Documents*, I, 251-252.

[71] Halifax *North-Carolina Journal*, January 7, 1805; *Farmer's Register*, VII (December, 1839), 729.

[72] *U. S. Census of 1860* (Agriculture), 105-111.

[73] *Ibid.* The Brunswick County rice crop for 1859 was 6,775,286 pounds.

[74] *Farmer's Register*, VIII (April, 1840), 243.

[75] *U. S. Census of 1860* (Agriculture), 105-111.

declined thereafter. In North Carolina, the decline did not set in until after 1860. In the censuses of 1839, 1849, and 1859 the state's share in the total national product was approximately one-fortieth, one-forty-third, and one-twenty-sixth, respectively.[76]

The principal varieties grown were the "Carolina Gold" and the "Carolina White," both of which were introduced to the state from South Carolina. The former was superior in quality and thought to be the first rice imported into the Coastal region for planting purposes. The latter was very similar to the gold rice, but, because of its tendency to shatter when being harvested late, was not so widely grown.[77] There was also a variety of rice which was adapted to culture in the uplands. Its growth, however, never did become widespread in North Carolina.[78]

In the low-country where it was grown, rice was highly esteemed as a food by both whites and blacks. Being nutritious and healthy, and capable of preparation in several kinds of dishes, rice became the most popular item in the diet of that region.[79] The quality of that grown along the Cape Fear was of the finest and producers in other sections bought it for use as seed.[80] The balance of the product, over and above that required for the next year's planting, was disposed of by factors located in New Bern, Wilmington, and Charleston.[81]

Because of the unhealthy and unfavorable circumstances under which rice was grown, perhaps, a greater portion of the crop was produced by slave labor than any other grown in the state.[82] Conditioned through generations of life in the hot, steamy jungles of Africa, the Negro slave was better equipped for life and work in the rice growing section than were white men.[83] Out of this situation grew a practice of "too general prevalence among planters of absenteeism," which meant

[76] L. C. Gray, *History of Agriculture*, II, 723.

[77] D. C. Heyward, *Seed From Madagascar*, 4, 8.

[78] *North Carolina Planter*, I (April, 1858), 118.

[79] Patrick H. Mell, "Rice Planting in the Agricultural Development of the South," *The South in the Building of the Nation*, V, 176.

[80] D. C. Heyward, *Seed From Madagascar*, 5.

[81] The best collection of materials bearing on the whole gamut of relationships between a rice planter and his factors is to be found in J. H. Easterby (ed.), *The South Carolina Rice Plantation*, 357-439. Charleston was the most important market for rice just as New Orleans was for cotton. There, in price per pound, rice reached its peak in 1817 at seven cents. In 1805 the price was next highest at six cents. The third highest was in 1855 when it brought five cents. The lowest price of the whole period was two cents which it brought in 1814. D. C. Heyward, *Seed From Madagascar*, 43 n.

[82] To those who worked in the rice-growing regions, the attacks of "bilious fever," which they had almost every year, were accepted as a matter of course. Edmund Ruffin said: "I cannot but consider the culture of rice as a curse to any region, as it must cause the production of malaria and its consequent diseases. . . ." *Farmer's Register*, VIII (April, 1840), 244.

[83] P. H. Mell, "Rice Planting in the Agricultural Development of the South," *The South in the Building of the Nation*, V, 175. For an excellent contemporary account of slave conditions on an ante-bellum rice plantation, see J. H. Easterby (ed.), *The South Carolina Rice Plantation*, 346-350.

that actual direction and control of operations during the greater part of the year devolved upon overseers.[84] The relationship, therefore, between master and slave was more impersonal, and, doubtless, the lot of the slave on the rice plantation was worse than that on any other kind of plantation.[85]

The methods used in rice growing changed but little from those employed in colonial times. Although animal drawn implements were introduced, the hand with the hoe continued to perform nearly all of the work.[86] This meant that the number of acres planted per field hand must be kept small. A contemporary said that "If the land is *well drained* and in *good order,* five acres of rice, and 1 or 1½ of provisions may easily be cultivated to the hand."[87]

The cultivable part of a typical rice plantation was divided into "tasks" by ditches, which usually crossed at right angles. Several "tasks" made a "field." The "fields" were numbered and contained from ten to twenty acres each. They were bordered by a large ditch and an embankment—the ditch serving both as an inlet and outlet for the larger canal, with which it connected. The flow of water from the canal to the ditch was controlled by flood gates.[88]

Preparation of the land for the next crop usually began in December when the stubble of the previous crop was burned, and the hands, with hoes, dug up the soil to a shallow depth. Between then and the beginning of planting, the hands were kept busy cleaning out the ditches, repairing weak places in the embankments of ditches and canals, and other maintenance work. The fields, of course, had to be drained and kept comparatively dry while this work was going on.[89]

[84] Robert F. W. Allston, of the Georgetown District in South Carolina, and one of the most prosperous and influential rice growers, found in a study of 111 rice plantations that only 51 of the owners actually resided on their plantations. "Of the major number," he wrote, "all were absent, and are habitually absent during the crop season, from some time in the month of May to November frost. Twelve of this number have their residence elsewhere." *North Carolina Farmer,* III (February, 1848), 193-194.

[85] E. Q. Hawk, *Economic History of the South,* 261; R. H. Taylor, *Slaveholding in North Carolina,* 15, 41-42.

[86] A drill plow was introduced about 1812, and, at first, excited the interest of rice growers, but was soon abandoned because "it required more minute attention and judgment than could be calculated on among the field laborers of that day." *DeBow's Review,* I (April, 1846), 339-340. A visitor to the rice plantation of Colonel T. D. Mears, of Brunswick County, in 1859, described as "one of the largest" on the Cape Fear, wrote: "No plows are used, the whole work being done with a broad short hoe and a long narrow one." *North Carolina Planter,* II (May, 1859), 131-132.

[87] *Southern Agriculturist,* IV (June, 1831), 292.

[88] *North Carolina Planter,* II (May, 1859), 131-132.

[89] *DeBows' Review,* IX (October, 1850), 421-424; *North Carolina Planter,* II (May, 1859), 131-132; D. C. Heyward, *Seed From Madagascar,* 28-31. R. F. W. Allston, of Georgetown, South Carolina, has an interesting article on the history of rice, methods in planting, total crops, sales, uses, *et cetera,* in *DeBow's Review,* I (April, 1846), 320-357. For a collection of articles by contemporaries on the subject of rice planting, see J. D. B. DeBow, *Industrial Resources,* II, 392-434.

Planting began about the tenth of March and sometimes continued until the tenth of June. Shallow trenches were dug with hoes, about six to ten inches apart, the rice dropped in, and the seed covered with a light covering of earth. When a field was planted, the water was turned on it to a shallow depth in what was called the "sprout flow." This was designed to help the seed to sprout and to kill young weeds and grass which did not thrive in water as did rice.[90]

The "open trench" system of planting came as a development in rice growing during the period. In this system, before planting, the seed rice was placed in water and then rolled in clay until it had collected a thin coating. Then it was planted in an open trench, which was not covered, and the sprout flow was turned on the field.[91] This precaution had to be taken to safeguard the seed against being eaten by the May birds which abounded in the Coastal region in the spring.[92]

The sprout flow was allowed to cover the fields four or five days, or until the seed sprouted, then it was withdrawn by opening the flood gates at low tide. When the rice had grown to the point that it could be seen in rows at a distance of about fifty yards, the fields were again flooded with what was called the "point" or "stretch" flow. This protected the young plant from the birds, softened the earth, and helped kill young weeds and grass. This flow covered the fields for from three to seven days, after which, it was drained off. By that time the rice had become three or four inches tall. It then entered upon a period of dry growth of about forty days, depending upon rainfall.[93]

During this time the crop was given two or more hoeings, or "weedings" as they were called, the principal purpose being to keep it as free as possible of weeds and grass. This was done by hoe-hands using the ordinary hoes.[94] The softness of the earth and the narrow rows prevented the use of animal drawn cultivators. When the rice had attained a height of about fifteen inches, the fields were again flooded with the "long" or "harvest" flow. This retarded the growth of weeds and grass and helped prevent the rice from being blown over. About two weeks prior to the beginning of the harvest the water was again drawn off the fields.[95]

Harvesting was done by hand sickle. A handful was grasped with the left hand, cut off, and laid atop the line of stubble. Other hands followed, collecting the rice into bundles or sheaves for removal from the field.[96]

[90] *DeBow's Review,* IX (October, 1850), 422; *North Carolina Planter,* II (May, 1859), 131-132; D. C. Heyward, *Seed From Madagascar, 27-37*; *Southern Agriculturist,* II (May, 1829), 193-197.

[91] *DeBow's Review,* IX (October, 1850), 423-424.

[92] D. C. Heyward, *Seed From Madagascar,* 28-31.

[93] *DeBow's Review,* IX (October, 1850), 422-423.

[94] *Ibid.,* 424-425.

[95] *Ibid.*

[96] *DeBow's Review,* IX (October, 1850), 424-425.

Although a threshing mill was introduced in the rice industry about 1811, a great part of the crop continued to be threshed out by the hand flail.[97] Seed for the next planting were never threshed by mill because of the great likelihood of injury, but were carefully threshed out by whipping the grain from the straw.[98] The grain was separated from the chaff by winnowing or fanning. After this, it was put through a process of "pounding," either in a pounding mill or by hand with a crude mortar and pestle, for the purpose of removing the outer hull. The grain was then ready for use as a food or for putting into barrels for shipment.[99]

Very little manure was used in the cultivation of rice and the land on which it was grown was not adapted to a rotation of crops. Generally the crop was planted year after year on the same land, and, when the product began to decrease, the land was sometimes rested for a year in natural growth. Around thirty bushels to the acre was considered a satisfactory crop.[100]

Among the scourges the rice planter had to contend with, "volunteer" rice was mentioned, perhaps, more than any other. This was the "renegade" grain that came up in the crop. Most of it shelled out before harvest time, but those grains which remained could not be cleaned by pounding and, therefore, reduced the quality of the grain.[101]

"May" or "rice" birds were a menace to the crops, too. These birds migrated north at about the same time the crop was being planted, and, unless guarded against, could severely damage the crop in a short time. At the time of harvest, birds again manaced the crop. Often, a slave armed with a gun was stationed in the rice fields to frighten them away.

Heavy tides and storms which beset the Coastal region, beginning in the late summer often damaged, and sometimes totally destroyed, the rice crops. One such occurred along the North Carolina coast on August 19, 1837, destroying all the water mills along the Cape Fear, except that at "Orton" plantation, and a great part of the rice crop for that year.[102]

[97] *Ibid.*, I (April, 1846), 340. The Josiah Collins estate on Lake Phelps, when advertised for sale in 1805, was said to have a "threshing machine in excellent order." Halifax *North-Carolina Journal,* June 7, 1805. "Orton" plantation, on the Cape Fear, was equipped with a water-operated "rice machine." U. B. Phillips (ed.), *Plantation and Frontier Documents,* I, 251-252. Colonel T. D. Mears, of Brunswick County, used steam driven machinery "of the best pattern," in cleaning his rice. *North Carolina Planter,* II (May, 1859), 131-132. See also, D. C. Heyward, *Seed From Madagascar,* 21-24.

[98] David Doar, *Rice and Rice Planting in the South Carolina Low Country,* 15-16.

[99] D. C. Heyward, *Seed From Madagascar,* 21-24.

[100] *North Carolina Planter,* II (May, 1859), 132; *DeBow's Review,* IX (October, 1850), 426; D. C. Heyward, *Seed From Madagascar,* 42.

[101] *DeBow's Review,* IX (October, 1850), 425-426; David Doar, *Rice and Rice Planting in the South Carolina Low Country,* 23.

[102] Tarboro *Press,* September 2, 1837. About once every ten or twelve years, these storms reached tropical proportions. Their destructiveness has been cited as

Although confined to a relatively small section of the state, rice growing continued through the ante-bellum period as an important agricultural interest. Stimulated by the movement for general improvement of farming in the state, which began in the late thirties, and the high prices being given for the grain, rice growing in North Carolina was on the increase in the 1850's. Evidence of this began to appear in the press. The editor of the *Newbernian,* in noting the quantity of rice being shipped from Newbern, remarked: "There is a large portion of land in this section well adapted to this article . . . our planters should put in a large seeding."[103] At the very end of the period, a Robeson County contributor surveyed the quality of the land in that section, and remarked: "I think the time is not far distant, when this entire country will be devoted to raising rice."[104] Whatever degree of realization this dream may have had, however, was cut short with the coming of the war and the consequent freeing of the slaves. Rice growing reached its peak in North Carolina in 1859. Since then the culture of the crop has been abandoned.

The other cereal crops, oats, rye, barley, and buckwheat, were decidedly minor crops throughout the ante-bellum period in North Carolina. On every farm and plantation, where planted at all, they were usually in competition with one or more "money" crops, such as, cotton, tobacco, or rice, and always with the more popular cereals, corn and wheat.[105] The result was that these minor cereals were usually relegated to the poorest lands on the farm, and were given even more careless and slovenly treatment than was noted as the common practice in wheat raising. Per acre yields, therefore, were very small and the tendency was to take acreage away from these crops when prospects were good for the more important crops.[106]

Oats which were more widely grown than the other minor cereals, suffered most in competition with the money crops. For example, in the decade of the fifties, while the production of cotton and tobacco was increased, the production of oats declined by more than one-third.[107]

---

one of the principal reasons South Carolina and Georgia, which produced a lion's share of the ante-bellum crop, finally abandoned the culture of rice. D. C. Heyward, *Seed From Madagascar,* 221.

[103] Quoted in *North Carolina Farmer,* III (March, 1848), 222.

[104] *North Carolina Planter,* I (April, 1858), 118.

[105] Oats generally sown on the worst land," wrote an Orange County reporter in 1850, "having been planted with corn the previous year: regarded an exhausting crop. Seed per acre, 4 to 5 pecks. Product in good ground, 30 to 40 bushels; common average, about 10 bushels." *Report of the Commissioner of Patents* (1850), 376. See also, T. F. Hunt, "The History of Cereal Farming in the South," *The South in the Building of the Nation,* V, 221.

[106] A Cabarrus County reporter wrote: ". . . the best ground must go for cotton, next best for corn, and the third quality for oats. But if oats had the ground cotton gets, the yield would be twenty-five and thirty bushels per acre." *Report of the Commissioner of Patents* (1852), 93-94.

[107] *U. S. Census of 1860* (Agriculture), 105-111.

Doubtless, ground was also lost to corn and wheat, both of which showed appreciable increases in production in the decade. Unfortunately for oats, the crop had also acquired the unwarranted reputation as a greater exhauster of the soil than wheat. The fallacy of this was attacked in the latter part of the period, but too late to bring oats into better repute with farmers generally.[108] Oats were used principally as feed for horses, and had only slight importance as a source of farm income.

Rye had a better reputation as a crop, but was not planted so generally. Used principally as a food for man, it was planted chiefly in the Piedmont, where the heaviest German settlement had been made, and, with whom, it was popular when used in bread making. Rye, too, was usually planted on the worst lands. However, the product showed a considerable increase in the latter part of the period.[109]

Barley and buckwheat were decidedly less important than either oats or rye. Most of that grown was in the Piedmont and Mountain sections where they were used as food for man. Barley production remained about constant in the latter part of the period. Buckwheat production in the 1850's, however, reflected the general progress of agriculture in the state by increasing over one hundred per cent.[110]

Of the cereals generally, ante-bellum North Carolina agriculturists, whether operating small farms or large plantations, sought in the first place to produce their own requirements. This in the main was substantially accomplished. In the second place, these crops were raised for outside sale. For some individuals, this was done profitably. In the culture of these crops, the better producers kept abreast of the times. Like cereal producers everywhere at that time, however, they suffered great losses to the diseases and insects which attacked their crops. To the great mass of farmers, these losses were accepted as "providential and unpreventable."[111] As for efficiency of tillage, the practices of ante-bellum North Carolina cereal producers ranged from the best to the worst, sometimes in adjoining patches. On most plantations and farms, the greater emphasis was placed on either tobacco or cotton. The cereals, therefore, were relegated to the category of "minor crops" and the poorer lands were usually assigned to them. Under such circumstances, when "seasons" came, the more important crop was worked first. The lesser crops had to wait. Such neglect helps explain the smallness of per acre yields in the state generally.

Cereal crops benefited also from the urge towards agricultural im-

[108] "Super-Phosphate" wrote that this fallacious idea had been "handed down to us, founded only in ignorance and blind credulity—a sort of superstitious belief in the simple 'say so' of those who have gone before us," and could only be exposed and removed by science. *North Carolina Planter,* I (February, 1858), 35.

[109] *U. S. Census of 1860* (Agriculture), 105-111.

[110] *U. S. Census of 1860* (Agriculture), 105-111.

[111] B. T. Galloway, "Plant Pathology; A Review of the Development of the Science in the United States," *Agricultural History,* II (April, 1928), 49-51.

provement which began to be felt in the state after about 1835. By the end of the period substantial progress was being made in the culture of the cereals, not only from the standpoint of improvements in the mode of culture, but also in the volume and quality of product. Changes in the volume of cereal production, for the last three census years, are shown in the following table.[112]

CEREAL PRODUCTION IN NORTH CAROLINA, 1840, 1850, AND 1860

| Item | 1840 | 1850 | 1860 |
|---|---|---|---|
| Corn, bu | 23,893,763 | 27,941,051 | 30,078,564 |
| Wheat, bu | 1,960,855 | 2,130,102 | 4,743,706 |
| Rice, lbs | 2,820,388 | 5,465,868 | 7,593,976 |
| Rye, bu | 213,971 | 229,563 | 436,865 |
| Oats, bu | 3,193,941 | 4,052,078 | 2,781,860 |
| Barley, bu | 3,574 | 2,735 | 3,445 |
| Buckwheat, bu | 15,391 | 16,704 | 35,924 |

[112] *Compendium of the Enumeration of the Inhabitants and Statistics of the United States as Obtained . . . from the Returns of the Sixth Census*, 176; *The Seventh Census of the United States: 1850. . . .*, 319-322; *U. S. Census of 1860* (Agriculture), 105-111.

CHAPTER VIII

# FRUITS, VEGETABLES, AND OTHER MINOR CROPS, 1820-1860

Fruit and vegetable growing in ante-bellum North Carolina, like most other features of its agriculture, presents a study in extremes. Where, on the one hand, a few individuals were growing products of fine quality and making significant contributions to the development of this type of agriculture, there is, on the other hand, much evidence that with most producers these crops suffered generally from carelessness and neglect. This was more the case with fruits than with vegetables.

Several factors must be borne in mind in a consideration of ante-bellum fruit and vegetable growing. Prior to 1860, these items attained only slight importance in the commercial life of the state. There was no large urban population to be fed, in fact, most dwellers in North Carolina's small towns still managed to keep their own gardens and orchards. For the greater part of the period, transportation facilities were such that it was impractical to ship perishable items unless first processed into such forms as dried fruit or brandy. There were no commercial canneries to take care of the surpluses. Fruits and vegetables were not so prominent in the diet then as now. Also, these crops, like all other minor crops, suffered from neglect when they came into competition with cotton and tobacco.[1]

By 1820, every fruit now grown in the state and some which are not now grown had been introduced. In the main, these were bought from northern and European nurseries for at that time there were no nurseries in North Carolina. Some varieties, particularly peaches, throve and produced fine fruit from the beginning. Others, particularly apples, pears, plums, and grapes did not adapt themselves so readily to the soil and showed the need of acclimitization. As a Releigh correspondent expressed it: "To be laconic—they have not done well. . . ."[2] Still others, such as the date and citrus fruits, did not thrive at all.

With the average grower, fruit trees received no further culture after they were set out. Naturally the quality of the fruit degenerated and the life of the orchard was cut short. This carelessness and inat-

[1] Andrew M. Soule, "Vegetables, Fruit and Nursery Products, and Truck Farming in the South," *The South in the Building of the Nation,* V, 236-241; L. C. Gray, *History of Agriculture,* II, 824 ff.

[2] This writer claimed to have been an "importer" of apple, pear, peach, cherry, and plum trees, since 1786, chiefly from the Messrs. Princes, of Flushing, Long Island. *American Farmer,* X (January, 1829), 339.

tention in the culture of fruits early attracted the wrath of the agricultural reformer. George W. Jeffreys wrote as follows on the subject:

> The feelings of a lover of improvement can scarcely be expressed, on observing the almost universal inattention paid to the greater number of our orchards, and that people who go to a considerable expense in planting and establishing them, afterwards leave them to the rude hand of nature; as if the art and ingenuity of man availed nothing, or that they merited no further care. . . .[3]

John Taylor, of Caroline, emphasized fencing of the orchard as a fundamental requirement before better fruit could be produced.[4]

Other individuals saw the need for reform and began experimenting with fruit trees. Benjamin Vestel, of Chatham County, for instance, reported to the West Cane Creek Agricultural Society that he had planted one hundred apple trees in 1819 but that they were "of a very indifferent and unthrifty growth; they were knotty, and inclined much to growing of sprouts from the body and roots. . . ." To correct this condition, he cut away the sprouts, topped and grafted the trees, and applied a white wash made of lime to the trunks of the trees; "except a few of them, which I left unlimed to prove the experiment. These few remain rough, sprouty and unthrifty; while those that were limed have cast off their rough coats of bark, have but little disposition to growing of sprouts, and are now thrifty and growing trees."[5]

From such experimentation, a body of information was gathered which was disseminated throughout the state by the agricultural societies and press. Doctor John F. Tompkins, editor of the *Farmer's Journal,* severely condemned the carelessness with which most fruit growers tended their orchards and urgently recommended more intelligent culture. "In our opinion," he wrote, "there is no branch of agriculture that would pay better in a short time than a well conducted orchard stocked with choice fruit, in North Carolina."[6] "Half a dozen choice, thrifty and productive fruit trees are worth more than one hundred poor ones, and, while the latter seem only a curse to the owner, the former will prove a source of continual and increasing profit."[7]

In this, as in other phases of the agricultural reform movement, the heaviest emphasis was given to the benefits that would flow from a wise selection of plant varieties, and a proper preparation of the soil, culture, and care of the growing trees. Under the extensive practices of the pioneers the tendencies were towards degeneracy of the trees and the

[3] (George W. Jeffreys), *Essays on Agriculture,* 144. Jeffreys wrote several essays on the subject. *Ibid.,* 144-157.

[4] From reproduction of *Arator* in *Farmer's Register,* VIII (December, 1840), 748.

[5] *American Farmer,* III (September, 1821), 207-208.

[6] *Farmer's Journal,* II (June, 1853), 84.

[7] *Ibid.,* 111.

production of inferior fruits. The aims of the reformers were towards intensive culture which, it was said, would produce abundant harvests of high quality from a smaller number of trees. As a "Fixed Fact" of agriculture, it was emphasized by the reformers that

> The *orchard,* to be productive of good, fair fruit, requires to be fed as much as does a field of grain. The soil of each requires that the substances abstracted by the crops should be restored. The soil should be kept clean, and open to the meliorating influences of the sun, the dews, the rain and the air,—the bark of the trees should be kept in a healthful condition, by scraping, when necessary, and by alkaline washes.[8]

It was soon found that cultivation, alone, would not guarantee fine fruits.[9] The plant variety had to be acclimatized. Also, it was discovered that some varieties produced better fruit in one section of the state than in another. To find the varieties best suited to his soil and location became the goal of every intelligent fruit grower. To meet the need posed by this situation, nurseries were first establishd in the state and several individuals, notably, Silas McDowell, of Macon County, Joshua Lindley, of Guilford County, and Sidney Weller, of Halifax County, won distinction for themselves for their work with apples, peaches, and grapes, respectively.

The Lindley Nursery, which was started on Cane Creek in Chatham County about 1826, but later moved to New Garden, in Guilford County, was, perhaps, the first nursery established in the state. Stock for this nursery was collected over a long period of time from the leading nurseries and collections in the United States. The trees were experimented with and tested for actual bearing before being offered for sale. Many varieties of apple, peach, cherry, plum, apricot, nectarine, and quince trees, grape vines, and strawberries were sold at nominal prices.[10] Lindley advertised his stock in a descriptive catalogue, in the farm journals, and by exhibition at fairs.[11] One hundred and twenty-seven varieties of peaches, alone, were tested for bearing and advertised in a descriptive list.[12]

Charles Mock, of Davidson County, was one of the first to establish a nursery in the state. He was the first to put out a catalogue advertising his product. His stock consisted chiefly of apple, pear, peach, plum, cherry, and nectarine trees; also two hundred varieties of roses and dahlias, which he offered to sell fifty per cent lower than "the same

[8] Reproduced from *American Farmer* in *Farmer's Journal,* I (August, 1852), 132.

[9] *Arator,* I (August, 1855), 148.

[10] *North Carolina Farmer,* I (February, 1846), 214-216.

[11] *Arator.* I (November, 1855), 256. This nursery exhibited over one hundred varieties of apples, described as "extra fine," at the North Carolina State Fair in 1858. *North Carolina Planter,* I (November, 1858), 347.

[12] This list was reproduced in *Carolina Cultivator,* II (April, 1856), 55-56; *ibid.,* (May, 1856), 74-77; *ibid.,* (June, 1856), 118-120; *ibid.,* (July, 1856), 149-152.

articles can be had at any other nursery in the Union," and promised delivery to any place within a hundred mile radius of Lexington.[13]

In 1843, S. W. Westbrook established the West Green Pomological Gardens and Nurseries near Greensboro. In 1852, he bought out the nursery of Charles Mock, of Davidson County, and moved the stock of trees to West Green. This business occupied about fifty acres, and, by 1858, had growing 500,000 trees of various sizes, including numerous varieties of apples, peaches, pears, cherries, plums, apricots, and nectarines. Also under development was the strawberry, which had been "made a specialty" of the nursery, and, of which, there were thirty varieties. The nursery possessed both American and European varieties of grapes of "almost every kind, in a fine state of cultivation. . . ."[14]

This nursery advertised its wares by displaying fruits, chiefly the apple, in local and national exhibitions. In 1856, a correspondent of *The Country Gentleman* described the exhibit of apples which Westbrook made in New York City as "an extensive and valuable collection of southern apples." There were, he wrote, "about fifty varieties handsomely grown, and furnishing a very favorable indication of the quality of the best southern varieties, the collection taken as a whole being quite equal, we think, to many collections of equal number, grown in the best northern apple districts."[15] This display won highest honors in the 1856 exhibit, and the nursery received orders from northern and western apple growers for trees.[16] At the National Pomological Convention held in New York City in 1858, the Westbrook exhibit of apples was again the "finest and largest" collection shown. A correspondent of the New York *Times* described the general display of fruits at this convention as "much superior to any ever before given in this city." "Prominent in the exhibition," he wrote, "is the most attractive display of apples we have ever seen—77 varieties by Westbrook and Mendenhall, of Greensboro, N. C."[17]

In the 1850's, Walter L. Steele, of Rockingham, began the establishment of a nursery. Stock was procured by selecting the best varieties of pears, apples, peaches, cherries, plums, apricots, strawberries, and grapes that could be had from American sources, and 10,000 additional pear seedlings were ordered from France. Steele attended the National

[13] *Farmer's Advocate,* III (January, 1842), 240. Mock's reputation as a nurseryman was described by one correspondent as "unsurpassed." *North Carolina Planter,* I (September, 1858), 275-276.

[14] *North Carolina Planter,* I (September, 1858), 275-276.

[15] Reproduced in *Carolina Cultivator,* II (January, 1857), 335. These varieties were named and described in *ibid.,* 335-336.

[16] From New York *Times* reprint in *North Carolina Planter,* I (October, 1858), 323.

[17] *Ibid.,* 325. The exhibit of this nursery at the North Carolina State Fair in this year consisted of over one hundred varieties of apples, sixteen of pears, five of peaches, two of quinces, several raspberries, and twenty-six strawberries. *North Carolina Planter,* I (November, 1858), 347.

Pomological Society Convention in 1859 and exhibited ten varieties of his apples as well as one of pears and scuppernong grapes. In announcing the opening of his nursery, Steele said: "It is our purpose to carry on the business honestly, without puffing or humbuggery, and to sell at *reasonable* and *uniform* prices."[18]

The increase in interest in the production of better fruits was further manifested by the organization of horticultural sections of agricultural societies, and the increase in the number of articles on fruit growing which appeared in the agricultural press. In 1858, the Southern Pomological Society was formed at a meeting held in Charlotte. Several North Carolinians played prominent roles in founding that body. The first two annual meetings were also held in Charlotte. A fruit exhibition was arranged and all interested in fruit growing were invited to attend.[19]

A movement for the formation of a state horticultural society, in connection with the North Carolina State Agricultural Society, came too late in the period to have any influence on fruit growing. There was a feeling among some fruit growers that the latter body had been too neglectful of their interests, and that "the spirit of carpology among our people" needed to be stimulated.[20]

The *North Carolina Planter,* which began publication at Raleigh, in 1858, devoted a section to horticultural subjects, and selected S. W. Westbrook, the nurseryman noted above, as its editor. Westbrook announced his intention of bringing forth a number of essays on orcharding under the following divisions: (1) preparation of the land; (2) laying out the orchard plot; (3) transplanting; (4) selection of trees; (5) mulching and its effects; (6) value of annual top dressings; (7) training and trimming; and, (8) care of the trees.[21] Every issue of this paper carried articles by the editor on one or more of these subjects, or by contributors from over the state. Silas McDowell, the outstanding apple producer in the state, contributed several articles.[22]

[18] J. G. deR. Hamilton (ed.), *Papers of Thomas Ruffin,* III, 96-97.

[19] *North Carolina Planter,* II (August, 1859), 242; *Edgecombe Farm Journal,* I (September, 1860), 2.

[20] J. G. deR. Hamilton (ed.), *Papers of Thomas Ruffin,* III, 96-97.

[21] *North Carolina Planter,* I (January, 1858), 26. The agricultural books and manuals, which appeared in this period, also devoted space to horticulture. R. L. Allen's *American Farm Book,* for example, which was published in New York in 1850, dealt with nearly every fruit now cultivated in the United States, giving information as to proper soils, planting, pruning, grafting, budding, selection of trees, selection of varieties, diseases, and gathering and preservation of the fruit.

[22] McDowell, who was "almost exclusively" engaged in pomology, and "particularly that branch of the science which embraces Winter Apples," made a complete failure in his efforts to produce winter apples from trees procured from northern nurseries. By grafting on to seedlings procured from the Indians, however, he produced several varieties of winter apples of outstanding quality. McDowell gave names to these varieties, had colored drawings made of the fruit, and wrote of his experiences in several farm papers. *North Carolina Planter,* I (April, 1858), 125-126; *Southern Cultivator,* IV (November, 1846), 170.

McDowell was one of the first to exploit the superior apple growing qualities of the Mountain region. A traveling correspondent wrote of Macon County: "In no section of the United States have we seen finer apples, and they are mostly from seedlings originally planted by the Indians."[23]

The diseases and pests to which the various fruit trees were subject was a favorite topic of individual contributors to the horticultural press, and they used the opportunity to publicize their remedies. Blight was the enemy of the pear tree most often mentioned. The borer and a disease called "yellows" called forth many recipes for the protection of the peach tree. Lice and blight attacked apple trees. Most of the remedies called for the application of the "medicine" either around the roots or to the trunk of the trees. No instance was found of any sort of spraying.[24] As an example of the ingenuity used in combating one of these enemies, Silas McDowell's recipe for killing lice was prepared as follows: "After filling a large pot with tobacco stems, we boiled them for an hour, and then boiled the ambere [*sic*] until it was very strong: into this we put brown sugar, salt, spirits of turpentine, aloes, sweet milk and lime, we do not recollect in what quantities." The resultant mixture was applied to the trunks of the trees with a brush, and, according to McDowell, they were "greatly improved" by the treatment.[25]

Undoubtedly, the average fruit grower who was interested in the improvement of his orchard, must have been bewildered by the great number of varieties of fruits from which he had to choose. This confusion was constantly added to by the practice of individual growers who would take a standard variety of a fruit, such as the "wine sap" apple, and by grafting and budding with other apples produce a new variety, to which, they gave a name and often too extravagant praise.

"Amateur," for one, emphasized that cultivation, alone, would not make certain varieties valuable, and that great care should be taken to select only those that were "flavored well and suited to our climate." "A bushel of Bullock's Pippins," he wrote "is worth a cart load of common apples."[26] He recommended, as well adapted for planting in North Carolina, seventeen particular varieties of apples, six of peaches, eleven of pears, six of cherries, and nine of plums.[27]

Much of the fruit was preserved, chiefly by drying, for winter use or

[23] *DeBow's Review,* XXVI (June, 1859), 706.

[24] *Arator,* I (May, 1855), 55; *ibid.,* I (December, 1855), 274, 279; *North Carolina Planter,* I (June, 1858), 195-196.

[25] *North Carolina Planter,* I (June, 1858), 195-196.

[26] *Arator,* I (August, 1855), 148.

[27] *Ibid.,* 148-149. Stephen A. Norfleet inventoried seventeen varieties of apples and ten of peaches on his plantation in Bertie County. It is interesting to note, that of his seventeen varieties of apples, only one had a name at all similar to the names of the seventeen varieties recommended by "Amateur," and not one of his ten kinds of peaches had a name similar to those named by that source. Norfleet Diary, undated, Norfleet Papers.

for sale.[28] Nearly every community had one or more distilleries where large quantities of fruit were used in the making of brandies. Livestock, too, consumed a generous portion of the fruit harvests since in most situations neither was fenced in. Although many questions may be raised concerning the quality of the fruits produced in the state, there seems little doubt that the supply was adequate. Supplementary to the cultivated sorts were abundant supplies of wild fruits, berries, and nuts. The presence of this wild supply, undoubtedly exercised a retarding influence on the cultivation of the domestic varieties.

In the development of the grape industry, ante-bellum North Carolina played an important and significant part. No less than five varieties, which became outstanding in the nation, originated in the state. These were the "Lincoln," "Isabella," "Catawba," "Scuppernong," and "Hickman."[29] Sub-varieties of these were also developed.

Although some questioned the origin of the others, the Catawba, which became a popular variety over a great part of the nation, was first discovered in Buncombe County in 1802.[30] The scuppernong originated in Tyrrell County and was given that name in compliment to James Blount, of the town of Scuppernong, "who first diffused a general knowledge of it."[31] Nevertheless, the North Carolina nativity of this grape was challenged. Sidney Weller, of Halifax County, admitted that there were many "spurious kinds of scuppernong in the South that tend to disgrace the genuine," and proceeded to give a lengthy description of the real scuppernong.[32]

Confusion grew out of the similarity of varieties in grapes as in the other fruits. For example, F. J. Kron, of Stanly County, a native of France and sometimes member of the faculty of the state university, sent six bottles of wine to the state fair in 1860 which he said was made from "the *Lenoir,* alias *Warren,* alias *Herbemont,* alias *Lincoln,* but as I believe in truth, a Gamas, grape from France."[33]

In addition to the cultivated sorts, wild varieties of grapes grew in the forests throughout the state. The "fox" grape was one of these.

[28] The building of railroads in the state stimulated trade in fruit and fruit products. See William W. Holden, *Address Delivered Before the Cumberland County Agricultural Society in 1859,* 18; reprint of article from Fayetteville *Observer* in *Southern Cultivator,* XVII (December, 1859), 369. In the absence of adequate transportation facilities, John Taylor considered the apple the only fruit "capable of producing sufficient profit and comfort, to be a considerable object to the farmer." From reprint of the *Arator* in *Farmer's Register,* VIII (December, 1840), 788.

[29] Moses Ashley Curtis, *Geological and Natural History Survey of North Carolina, Part III, Botany,* 112-114.

[30] See *Western Agriculturists in Southern Cultivator,* XIV (August, 1856), 256; and *Cozzen's Wine Press* in *Farmer's Journal,* III (September, 1854), 185.

[31] *American Farmer,* III (January, 1822), 332. See also the Raleigh *Star,* March 15, 1833.

[32] *DeBow's Review,* IX (October, 1850), 379-381.

[33] J. G. deR. Hamilton (ed.), *Papers of Thomas Ruffin,* III, 98.

Another, similar to the fox grape, was discovered in Rutherford County which, it was said, ripened at "fodder pulling time," never rotted, but continued round on the vine until frost and then dried up.[34] Once the location of such grapes became known, they were gathered "about fodder pulling time."

The production of grapes was among the first features of agriculture in North Carolina to receive benefits from state aid. Procurement and distribution of grape vines was one of the first activities undertaken by the North Carolina Board of Agriculture. This body was superseded in the early 1820's by the original state agricultural society, which made annual distributions of vines throughout the remainder of the decade. Eleven different varieties were given out to interested parties by that body in 1830.[35]

Of the great number of varieties of grapes grown in the state, the scuppernong was perhaps the most popular and certainly the most written about. It thrives best in sandy soils with sand or sand-clay subsoils. This type of soil abounded in the Coastal region of the state, and was very little valued for agricultural purposes. There were no insect enemies or diseases that attacked the vine or fruit. The vines were of hardy growth, required little or no cultivation, although they responded well to fertilization, and produced best results when allowed to grow freely without much or any pruning. The fruit was delicious to eat and produced a very high quality of wine. The vines were sure bearers and age seemed to have no effect on them. They bloomed in early June and the grapes ripened in September.[36]

The usual manner of cultivating the scuppernong was to plant the vines about twenty feet apart, and train them on an open horizontal frame or scaffold. This was supported by posts about seven feet above the ground. The frame was added to from time to time as the vine grew in age and spread.[37] About sixty bushels of grapes from one of these vines was considered a normal harvest, and most growers found it necessary to cultivate only one.[38] Once the vine started bearing, usually about three years after planting, no more cultivation was given it. The roots were of shallow growth and it was considered best not to disturb them.

As to the culture of the vineyard and the varieties of grapes to be grown, considerable differences of opinion arose among American grape

[34] J. G. deR. Hamilton (ed.), *Papers of Thomas Ruffin*, II, 321. One writer said: "Our woods are full of the grape vines. They ascend to the very tops of the tallest trees, and there hang to the elements their rich fruit in beautiful clusters." *American Farmer*, XII (April, 1830), 27.

[35] For description of these, see reprint from Raleigh *Register* in *American Farmer*, XI (March, 1830), 411-412.

[36] R. C. Cool, *The Scuppernong Grape: Its Growth and Care Under Vineyard Conditions*, 5-15.

[37] *Farmer's Register*, VII (November, 1839), 702.

[38] *American Farmer*, IV (December, 1822), 295.

growers. The principal points of difference grew out of the fact that European varieties, when transplanted to American soil and given comparable care and attention, did not produce comparable results. The difference was particularly noticeable in the quality of the wine—the American wine being inferior. This led to considerable discussion, in which North Carolinians played a leading part, and to some discouragement of American growers. For example, Nicholas Longworth, of Ohio, and leading vintner in the United States, experimented with "a great number of foreign vines" from all sections of Europe over a period of thirty-five years and finally came to the conclusion that not one of them was suited to "our climate."[39] Joseph Togno, of Wilmington, challenged this conclusion as being too broad and vague when applied to the United States, which contained such a diversity of climates, and urged grape growers to try all varieties of vines and adopt only those suited to the particular locality.

> Let each act upon this plan, and independently of each other, and then each will find out in the course of time, what is best for each of us. This is true experimental philosophy; this is not putting the extinguisher over the minds of others, and bar the efforts of generations to come, who no doubt will be made wiser by our present trials, and even blunders.[40]

Togno followed his own advice by conducting numerous experiments with foreign vines. These, of course, suffered some damage in transportation, and, he was able to save only a small number "out of many thousands, constituting several hundred varieties," which he had imported. Togno considered it highly important that the best native stock be crossed with the European varieties which he did by grafting. "The European vines marry very readily with our native vines by grafts," he observed, "with the exception of the scuppernong and its varieties, which sturdily refuse all union with these outlandish strangers."[41]

F. J. Kron, who had had some experience with winemaking in France and who imported large numbers of French vines for his vineyard in Stanly County, thought the American varieties of wine grapes were deficient in saccharine content in comparison with European grapes, and, therefore, required the addition of sugar before quality wines could be produced. He found that the Lenoir grape, which was widely grown in the state, had all the necessary qualities of a good wine

[39] Reprint from *Daily Cincinnati Commercial* in *Farmer's Journal,* I (July, 1852), 103-104.

[40] Reprint from *Daily Cincinnati Commercial* in *Farmer's Journal,* I (July, 1852), 103-104. Togno, who prepared a report on grape culture in North Carolina for the Patent Office in 1851, was described by the Wilmington *Herald* as "a gentleman of extensive research and practical knowledge upon the subject of the grape and its varieties. . . ." *Report of the Commissioner of Patents* (1851), 48. This article is contained in *ibid.,* 48-51.

[41] *Farmer's Journal,* I (October, 1852), 193-195.

grape except sugar. ". . . only give it sugar," he wrote, "one or two pounds per gallon before fermentation and a wine can be made from it that will keep without the addition of alcohol, acquire a fine aroma in the course of 12 months, exhilerate the well and roborate [*sic*] the sick."[42]

Sidney Weller, of Halifax County, the outstanding grape producer in the state, after much observation and experimentation with foreign vines finally gave them up entirely in favor of the domestic varieties.[43] Using a native grape, he evolved a variety which he named "Weller's Halifax," and, for which, he asked premium prices. Weller vigorously opposed the European practice of close trimming and used what he called an "American system," which was, in brief, to permit the vines to grow freely on arbors and scaffolds with a minimum of pruning, which, he said, would not keep them "low or humble."[44] In the selection of varieties, culture of grapes, and in wine making (Weller also had an "American System" of wine making), Weller strongly and repeatedly urged that "America must throw herself, in this matter and others, on her own resources; and at most only take hints from foreign men and books, on subjects of agriculture. And not servilely *ape* foreigners in things that experience proves our peculiar climate and circumstances render inexpedient or destructive."[45]

Weller established a nursery in Halifax which was principally devoted to the propagation of grape vines, but also included apples, peaches, pears, "&c., the choicest sorts, and all selected with reference to adaptedness to southern climes."[46] The fruit trees were grafted and budded and sold for from twenty to fifty cents each "according to sorts, size, and scarcity in the market." He also sold from his nursery ornamental trees, strawberries, rhubarb, apricots, "fig plumbs," and various garden seed.[47] The vines and trees were boxed for shipment in damp saw dust or moss and withstood transportation with only small loss.[48] Weller maintained a large scuppernong vineyard in addition to the nursery. From these vines, he made in a normal year an average of

[42] J. G. deR. Hamilton (ed.), *Papers of Thomas Ruffin,* III, 98.

[43] *Farmer's Advocate,* III (January, 1842), 227. James Dunwoody Brownson DeBow called Weller "a most practical man" and "one of the most successful vintners in the whole south—we might say in the Union." *DeBow's Review,* VIII (March, 1850), 245.

[44] *Farmer's Advocate,* III (January, 1842), 227-228; *Farmer's Register,* IV (December, 1836), 459-460; Raleigh *Standard,* June 25, 1845. Weller did not discover his so-called "system," but he did more to encourage abandonment of European modes of culture than any one. *American Farmer,* XII (April, 1830), 27.

[45] *Farmer's Advocate,* III (January, 1842), 228-229.

[46] *Ibid.,* 230.

[47] *Ibid.,* III (April, 1842), 317-318.

[48] *North Carolina Farmer,* I (October, 1845), 103.

forty barrels of wine besides the quantities of grapes eaten or carried away by visitors.[49]

Weller was a prolific writer on various aspects of the agricultural situation, particularly on the subject of grape culture. He subscribed to all of the leading farm journals and contributed many articles of real merit to their pages.[50] He was a propagandist not only for his "American Systems" of grape culture and wine making, but also for general improvement in farming practices. He was an alert business man and made considerable profit out of his operations.

North Carolina ranked first, fifth, and sixth in wine production in the Union according to the censuses of 1840, 1850, and 1860, respectively. Explanation for this high standing may be attributed almost entirely to the superior wine making qualities of the scuppernong. Unlike most American varieties, this grape was found by analysis and practical experience to contain saccharine matter equal to the best French grapes.[51] Real scuppernong wine was so good, in fact, that cheap imitations were made and sold at premium prices. One correspondent noted that much "so called" scuppernong wine was made in Raleigh of "honey and other ingredients."[52] Edmund Ruffin said that most of what was called scuppernong wine was not wine at all, but a liquor produced by improper fermentation.[53] Another observer wrote: ". . . the inhabitants, after their rude and careless manner . . . do not suffer it to ferment, but add new apple brandy, to the juice just from the press. At the 3d day's fermentation, it is excellent Champaign, but few bottles can hold it in this climate."[54]

Undoubtedly, most of the wine made in the state was carelessly done and of poor quality. Described as an item "of daily consumption," it was used as soon as made and received none of the benefits of aging.[55]

Near 1860, there was some indication that wine making was being done more scientifically and carefully. Ebenezer Emmons urged that an analysis be made of the juices of wine grapes in order to determine a proper balance of ingredients to be used in the wine-making process. Good wine could only be produced by proper fermentation, a fact, he said, which many people did not know.[56]

[49] *DeBow's Review,* IV (November, 1847), 311. In 1849, Weller's wine production had reached fifty barrels a year and he claimed to be second only to Nicholas Longworth, of Ohio, in the whole country. *Ibid.,* VII (July, 1849), 59.

[50] His best articles on this subject were "Outline of an American Vineyard From Its Beginning," *North Carolina Farmer,* I (October, 1845), 104-105; "The Cost of Southern Vineyards," *DeBow's Review,* VIII (March, 1850), 245-249; "True Secret of American Wine Making," *ibid.,* VII (July, 1849), 59-62.

[51] Daniel Moreau Barringer, *Address Delivered Before the Mecklenburg Agricultural Society*, 18.

[52] *American Farmer,* III (January, 1822), 332-333.

[53] *Farmer's Register,* VIII (April, 1840), 253.

[54] *American Farmer,* IV (December, 1822), 295.

[55] *Ibid.,* XII (April, 1830), 27; *ibid.,* III (January, 1822), 332-333.

[56] *Legislative Document No. 26* (1860-1861), 4-5.

Grape culture and wine-making were encouraged in order to combat "*the great national sin of intemperance.*"[57] In 1859, a 3,000 vine vineyard was established on Roanoke Island containing Catawbas, Diannas, Concords, Isabellas, Malagas, Herbemonts, and Scuppernongs. The fruit from this vineyard was intended for the manufacture of wines and brandies.[58]

Although grape culture in ante-bellum North Carolina, from the financial standpoint, never proved of great importance to producers generally, it did provide considerable enjoyment for the people. Perhaps, from no other source was such great profit received for the expense and effort expended.

Except on the better farms, the vegetable garden, where it existed at all, offered very little in the way of variety and presented a sorely neglected spectable in the early 1820's.[59] The reasons for this are fairly obvious. Meat, of which the diet overwhelmingly consisted, was present in abundance—whether of the wild or domesticated sorts. From the dietary standpoint, vegetables were only slightly appreciated. Then too, with very little effort, an abundance of wild fruits, berries, and nuts could be procured from the forests and waste lands. As further obstacles to gardening, it was difficult to secure good seed, to safeguard the garden from destruction by foraging animals, and to preserve surpluses of produce except by drying.[60] With those people who lived on the frontier or who contemplated emigration, the gardens were particularly scanty and neglected.

Agricultural leaders emphasized the improvement of gardens as one feature of the movement for the improvement of agriculture generally. For example, Paul C. Cameron admonished the Orange county Agricultural Society to

> Surround your dwellings by fruitful and well kept gardens. A patch of cabbage, six rows of Irish potatoes, and three rows of onions, do not constitute a farmer's garden. Use only the best seed, and plant only the best fruit. It is as easy to cultivate the Vergalue and the Bartlett, as the Choke Pear; the Green Gage, as the Horse Plum. No man is better entitled to all the good fruits of the earth, than he who tills it. And Lord Bacon declares,

[57] D. M. Barringer, *Address Delivered Before the Mecklenburg Agricultural Society,* 18-22.

[58] *North Carolina Planter,* II (June, 1859), 190.

[59] A traveler through the western Piedmont in 1802 observed that "The major part of the inhabitants do not even cultivate vegetables; and out of twenty there is scarcely one of them that plants a small bed of cabbages; and when they do, it is in the same field as the Indian wheat." R. G. Thwaites (ed.), *Travels West of the Alleganies Made . . . in 1802 by F. A. Michaux. . . . ,* 299.

[60] One woman of Perquimans County, who claimed that the management of the garden belonged to her sex, complained of the difficulty of getting good seed and urged agricultural societies to promote the establishment of nurseries and arrange to distribute seed and plants. "North Carolina," she wrote, "labours under the disadvantage, of being more remote from intercourse and improvement, than her sister states." *American Farmer,* IV (October, 1822), 223.

of the garden, 'that it affords the purest of human pleasures, the greatest refreshment to the spirit of man, without which, buildings and palaces are but gross handiworks.' "[61]

In this, as in other phases of the reform movement, editors of farm journals urged the adoption of more intelligent practices and threw their pages open to contributors of articles on the subject. "Pass through the garden once a day, at least," advised the editor of the *Farmer's Journal;* "give it an hour in the morning, and another in the evening, if possible; no part of the farm will pay you better than the garden crops."[62] The editor of the *North Carolina Planter* recommended that gardens be well cleaned of weeds, before they went to seed, manured, and broken up in the fall. The fall plowing would expose many insects to destruction by the frost, help keep the soil loose, and cut down on the growth of weeds and grass; but it should be followed by a spring plowing before planting. "Most persons have a place which they dignify by the name of garden," he wrote, "but at this season of the year they more resemble a cane brake or wilderness."[63] Advice was given as to the best type of soil and modes of culture of such vegetables as asparagus, beets, cabbage, carrots, celery, radishes, tomatoes, and turnips.[64]

Mercantile establishments began procuring and advertising garden seed as standard items of stock. As an example, William Hunter, of Charlotte, advertised the following sorts of fresh seed for sale: nine varieties of cabbage, five of lettuce, three of carrots, three of beets, five of turnips, three of cucumbers, two of parsnips, five of garden peas, four of beans, leek, parsley, "ockra," cayenne pepper, onion, celery, sage, musk melon, and sun flower.[65] Doubtless many farmers and housewives were not at all acquainted with some of these vegetables, and their experiments, both in the planting and in the preparation for eating, must have been ill-guided. To assist the farmer in the purchase of his seed, information was published as to how much seed of the various vegetables would be required to produce a certain number of plants or to plant a certain quantity of ground.[66]

The role of the wife as the manager of the garden, even on the most well-to-do plantations, is reflected in the papers of Judge Thomas Ruffin. He wrote his daughter: "I am afraid too that your Mama's Yams may have been frost bitten. If this be so, I would advise her to get Mr. Moring or his stage driver to buy her a bushel in Raleigh and to have them buried immediately."[67] On another occasion, Ruffin's

[61] *Arator,* I (June, 1855), 73.
[62] *Farmer's Journal,* II (August, 1853), 138.
[63] *North Carolina Planter,* I (January, 1858), 30.
[64] For example, see reprint from *Real Estate Register* in *North Carolina Planter,* I (January, 1858), 30-32.
[65] Salisbury *Western Carolinian,* February 14, 1835.
[66] *Arator,* I (August, 1855), 150.
[67] J. G. deR. Hamilton (ed.), *Papers of Thomas Ruffin,* I, 444.

son wrote him: "Mama complains greviously of her garden. . . . She says it is 'A wilderness of weeds,' and declares that she will never again be cheated of the labour of her gardener, by an overseer."[68]

Of all vegetables grown, the sweet potato was produced in larger quantity and was, by far, more beneficial than any other. It was grown in each of the state's eighty-six counties in 1860. Only seven of these counties showed a production of less than 10,000 bushels each, and forty counties produced more than 50,000 bushels. Duplin, Sampson, Wake, Johnston, and Edgecombe, all eastern counties, led in the order named with an average production of about 250,000 bushels.[69]

The sweet potato was raised almost entirely for home consumption, being highly prized as a food for both man and cattle. This was the case for high and low alike. Judge Thomas Ruffin, when threatened with the loss of his crop by frost, wrote: "I shall be very sorry to lose the crop of them; for I make so much of my winter's living on them, that the want of potatoes would be worse than the want of bread."[70] James W. Jeffreys, of Person County, who started the practice of planting the potato's sprouts or "slips" rather than the potato itself, declared the sweet potato "one of the most valuable [crops] which a farmer can raise, and as food for stock it is especially so, and every attention ought to be paid to the culture and propagation of that useful esculent."[71]

As manures and the other features of better farming were applied to the production of the crop, some tremendous yields were reported. One Moore County farmer grew 816 bushels on one acre.[72] Ebenezer Pettigrew, of Tyrrell County, produced one potato which weighed eleven and one-quarter pounds, was two feet long, and nineteen inches in circumference, besides, "several others of nearly the same size."[73] Successful potato growers began preparing articles for the farm journals describing their methods.[74]

The problem of preserving the potato against destruction by frost

[68] *Ibid.*, II, 119.

[69] *U. S. Census of 1860* (Agriculture), 105-111. The sweet potato thrives best on well-drained sandy lands. Hence, the East, having more soils of this type, dominated the state in the production of this vegetable. Hugh Hammond Bennett, *The Soils and Agriculture of the Southern States,* 27. In 1860, North Carolina stood second in relative rank in the Union in the production of sweet potatoes, having advanced from third place in 1850. *U. S. Census of 1860* (Agriculture), xxix.

[70] J. G. deR. Hamilton (ed.), *Papers of Thomas Ruffin,* I, 444.

[71] *American Farmer,* XI (April, 1829), 33. Jeffreys referred to his experiment with potato sprouts as "novel to me and unheard of." In planting in this manner, the producer required far less seed. The potatoes were bedded for sprouting and then the slips were pulled away from the mother potato and transplanted into the fields prepared for them.

[72] *Arator,* I (January, 1856), 307.

[73] *American Farmer,* VIII (December, 1826), 293.

[74] For examples, see article by Daniel Goggins to the agricultural society at Newberry, South Carolina, in *DeBow's Review,* XXV (October, 1858), 461-462; R. L. Allen, *American Farm Book,* 172-174.

and rotting was one that attracted considerable attention. Two methods of saving them were evolved. One system called for letting the potatoes remain in the ground until the vines were killed by the frost. Then, the potatoes were dug and sorted, putting the bruised ones aside for early consumption, and the good ones in "hills" prepared in the field or near the house for their safe-keeping.

The potato hills were made by throwing up the dirt into a mound and then leveling it off so that the top of the mound was slightly above the level of the ground roundabout. A large quantity of dry litter, usually wheat straw or pine straw, was then spread out on the mound. Then, from 50 to 100 bushels of the unbruised potatoes were piled atop the litter. The pile was then covered with straw to the depth of a foot or more and cornstalks and potato vines thrown on top to hold the straw down. Finally, he whole mass was covered by dirt to a depth of from four to six inches. A ditch was made around the hill to carry away water. The straw in the hill absorbed moisture which otherwise would cause the potato to rot. The covering of straw, cornstalks, and earth protected the hill against "rain and other carnalities."[75] This proved to be a successful mode of keeping sweet potatoes, and, with modifications by individual farmers, gained widespread use.

The other method consisted of spreading out the unbruised potatoes in a cellar or storage place which had a fairly even temperature, no draughts, was dry, and which protected the potatoes against sunlight and moisture. According to Doctor Gideon C. Marchant, of Indian Town,

> The whole secret consists in placing them away without bruising them—in keeping them at a proper temperature, in a large room, freely ventilated—and in preventing a current of air from passing through or over them.
>
> My cellar is never clear of potatoes.—I have kept them eighteen months as sound, and nearly as heavy as they were when taken from the earth.[76]

Although not grown on nearly so large a scale or in as much demand as a food, Irish potatoes were also produced in every county in the state. Only twelve counties out of eighty-six, in 1860, produced less than 5,000 bushels each. Buncombe, Guilford, and Randolph, in that order, led the state in production, averaging about 24,000 bushels each.[77] Near 1860, Irish potatoes were rapidly coming into prominence as a food. The increase in production between 1840 and 1860 was about one hundred per cent.[78] Had transportation facilities been better, the increase would, undoubtedly, have been greater.[79]

[75] This description of the potato hill was provided by "A Farmer," of Chatham County, in *Carolina Cultivator,* II (November, 1856), 279-280.

[76] *Farmer's Journal,* I (October, 1852), 210.

[77] *U. S. Census of 1860* (Agriculture), 105-111.

[78] *Ibid.*

[79] "Suwannano," of Buncombe County, who claimed he often produced as much

There were two principal methods used in planting Irish potatoes—in ridges or on the level. Each had its advocates and numerous articles were written favoring one or the other.[80] Per acre yields increased tremendously as manures and more thorough tillage were given the crop. Production of five hundred bushels per acre was not uncommon.[81]

Saving the Irish potato crop presented about the same problems as the sweet potato. Generally they were stored in a well ventilated place where they would be protected against moisture, sunlight, and freezing temperatures. An Orange County farmer said the secret in saving Irish potatoes was to dig them when the ground was dry and the sky overcast. "The old crop may be kept in good order for the table till the new one comes in," he wrote, *"provided that the sun be not allowed to look at them."*[82]

In the production of peas and beans, North Carolina retained the position of leadership she had attained in colonial times. In 1850 and 1860 the state held relative rank in the Union of first and second place, respectively, in the production of these crops.[83] These crops, too, were grown all over the state. Only nineteen of the state's eighty-six counties, in 1860, had a production of less than 5,000 bushels each. Wayne, Edgecombe, Onslow, Pitt, Bertie, and Sampson, all eastern counties, led in production in the order listed.[84]

These crops were grown not only in gardens, but extensively intertilled with other crops, particularly corn. Their popularity was increased by the agricultural reformers who generously praised them as soil builders.[85]

As items of trade and commerce, peas and beans never regained the prominence they had during the late colonial period, and were used almost entirely as subsistence crops for both man and beast. Peas, particularly, were planted for use in fattening hogs. For example, Stephen A. Norfleet, of Bertie County, raised enough peas on his plantation to fatten about three hundred hogs annually.[86]

Tomatoes or "love-apples," as they were sometimes called, began to be cultivated and used more widely. They had been grown for many years "for ornament and curiosity only." Tomatoes were eaten uncooked, pickled, preserved, and in ketchup. The varieties grown ranged in color from a bright yellow to a fine red, and in size from "about the

as five hundred bushels of Irish potatoes per acre, wrote that if he had the means of transportation he could make money raising potatoes at from twelve and one-half to eighteen cents per bushel. *Carolina Planter,* I (February, 1840), 27.

[80] *North Carolina Planter,* I (February, 1858), 56-57; *DeBow's Review,* XXV (October, 1858), 462-463; R. L. Allen, *American Farm Book* 167-172.

[81] *Carolina Planter,* I (February, 1840), 27; *Arator,* I (January, 1856), 307.

[82] Hillsboro *Recorder,* April 23, 1846.

[83] *U. S. Census of 1860* (Agriculture), xxix ff.

[84] *Ibid.,* 105-111.

[85] *American Farmer,* X (November, 1828), 265.

[86] Norfleet Diary, *passim,* Norfleet Papers.

size of a cherry" to some that "measured twelve inches in circumference."[87] Long existent prejudice against this vegetable, however, growing out of the suspicion that it was poison, prevented it from becoming popular, particularly with the mass of people.

North Carolina shared somewhat in the general enthusiasm felt for the Jerusalem artichoke, which was introduced into the state. Sidney Weller was one of those who advocated its culture. He estimated that his lands would produce from 800 to 1,200 bushels per acre and that the crop would grow anywhere the Irish potato could be grown. He recommended the crop for the following reasons: "1. As a *substitute* for the *Irish Potatoe* where it is no longer safe to cultivate on account of its destructive disease. 2. As a most profitable crop that can be made for fattening and store feeding swine. 3. As a most important auxiliary renovator of the soil where grown."[88] Weller and others prepared articles covering modes of planting and use of this vegetable, however, there is no evidence that its culture became widespread or important.[89]

The peanut or ground pea was another new crop introduced into the state.[90] Its culture did not spread rapidly at first, but during the 1850's, as its qualities became better known, agricultural leaders began to recommend it.[91] It was found that the soils which produced peanuts best were the light sandy soils which were of little value in the production of corn or cotton. One man could tend twelve acres in peanuts. Per acre yields ranged from thirty to one hundred bushels, and, in the 1850's, the price per bushel ranged from ninety cents to one dollar and a half with the demand constantly on the increase. One correspondent wrote of this crop in 1860: "It is decidedly the most profitable crop grown in the State in proportion to the number of hands required in tending it, and the amount of labor necessary to its perfect cultivation. But it has been almost entirely restricted to the sea-coast."[92]

The peanut was found to be a superior food for fattening hogs and much of the crop was used for that purpose. "I have grown them three years," wrote a contributor to the *Southern Cultivator*, "and must say that I have never found any thing to answer so good a purpose in fattening a hog, at the same time keeping the corn in the crib. . . ."[93]

An increasingly large quantity of the peanut crop was shipped to the northern markets, principally through Wilmington. The export in 1850 totalled about 50,000 bushels annually, but reached about 200,000

[87] *American Farmer*, IV (October, 1822), 218-219.

[88] *North Carolina Farmer*, II (January, 1847), 169.

[89] *North Carolina Farmer*, II (January, 1847), 169-171; reprint of article by Nathaniel Mason, of Northampton County, from *Southern Planter* in *ibid.*, II (February, 1847), 209; R. L. Allen, *American Farm Book*, 186.

[90] *North Carolina Planter*, III (November, 1860), 366.

[91] *Arator*, I (June, 1855), 71; *Executive Document No. 13* (1852), 19-20.

[92] *North Carolina Planter*, III (November, 1860), 366.

[93] Quoted in *North Carolina Farmer*, III (May, 1848), 285.

bushels by 1860.[94] In the latter year, an enthusiastic grower wrote: "This product bids fair to become a staple one in the state; the demand for it is constantly on the increase, the supply does not keep pace with it, and hence it will become so profitable that many will be induced to go into it."[95] The coming of the war, however, interfered with the natural development of the trend towards this crop and retarded the arrival of the peanut as one of the state's important commercial products.

Turnips, perhaps, received more persistent promotion from agricultural leaders than any other minor crop. They were recommended not only as food for man and livestock, but also, as valuable in resting lands used in the production of the more exhaustive crops. George W. Jeffreys was one of the crop's leading proponents and published an essay on the turnip's culture and uses.[96]

Turnips were principally used as winter food for livestock, usually being left in the ground until needed. They were fed either raw or cooked with corn or some other grain.[97] An Iredell County farmer expressed the high esteem in which the crop was held by some as follows: "I consider turnips are more profitable for a farmer than any other crop he could plant or sow on the same piece of ground."[98]

Of the other minor crops which received some attention from North Carolina agriculturists, hemp, flax, indigo, castor beans, sorghum, and sugar cane may be mentioned. Although none of these attained importance in the commercial life of the state, hemp and flax, perhaps, were grown more continuously than any of the other. Flax-seed, particularly, were usually listed in the "Prices Current" column of state papers.[99] Indigo never regained even the slight importance it had in colonial times, yet, some of it was grown in the state. About 1830, interest in its culture flared up in the Albemarle Sound area but soon died out.[100] Sorghum and sugar cane were experimented with chiefly in efforts to supply the home demand for molasses.[101] The census of 1860 indicated a great increase in the production of that item. It was made from sorghum almost entirely.[102]

[94] Reprint from Wilmington *Aurora* in Salisbury *Carolina Watchman*, January 3, 1850; *North Carolina Planter*, III (November, 1860), 366.

[95] *North Carolina Planter*, III (November, 1860), 366.

[96] (George W. Jeffreys), *Essays on Agriculture*, 139-143.

[97] *American Farmer*, II (December, 1820), 311; *North Carolina Farmer*, I (April, 1846), 250-251; R. L. Allen, *American Farm Book*, 175-178.

[98] *Report of the Commissioner of Patents* (1850), 369-370.

[99] Thomas Brent Moore, "Southern Hemp and Flax Production," *The South in the Building of the Nation*, V, 231 ff. Flax and flaxseed production declined about fifty per cent between 1850 and 1860. *U. S. Census of 1860* (Agriculture), 184-191.

[100] *American Farmer*, XI (December, 1829), 300.

[101] *Ibid.*, XII (April, 1830), 27; *Tarboro Southerner*, September 12, 1857; *North Carolina Farmer*, I (June, 1845), 17; J. G. deR. Hamilton (ed.), *Papers of Thomas Ruffin*, II, 534, 539-540; *North Carolina Planter*, I (January, 1859), 6-9.

[102] *U. S. Census of 1860* (Agriculture), 184-191.

North Carolinians participated to some extent in the numerous agricultural "manias," "fevers," or "crazes," which swept over the nation in the ante-bellum period. Merino sheep, Berkshire hogs, mammoth bullocks, Rohan potatoes, Baden corn, and mulberry trees and silk culture, were each the object of wild speculative build-up uniformly characterized by extravagant claims, intense excitement, disillusionment, and final collapse.[103] Usually agricultural leaders and editors of farm papers threw their influence behind these crazes in enthusiastic endorsements. Such promotion was referred to as "puffery," and once the craze subsided, it was labelled a "humbug."[104]

Undoubtedly the craze which attracted the most interest in the state was that having to do with the culture of the Chinese mulberry tree, or *morus multicaulis* as it was most often called, for the feeding of silk worms and the production of raw silk. First signs of this development began to appear in the state in the late 1820's.[105] Interest grew but slowly to about 1836, then rapidly rose to fever pitch in the late 1830's and early 1840's, and then collapsed resoundingly in 1842.

Several individuals and factors contributed to the contagion of this fever. It may be seen that the period of greatest interest in this crop coincides with the period when cotton production in North Carolina reached its point of lowest decline. Also, the idea grew that a silk industry would provide employment for paupers and dependent persons.[106] The state encouraged the culture and manufacture of silk by passing an act offering incorporation to any six or more citizens in a congressional district who should band together for that purpose.[107] The cause received its greatest boost, however, from the advocacy of certain individuals. Ebenezer Pettigrew and Josiah Collins of Tyrrell County, Robert H. Morison, President of Davidson College, and Sidney Weller, of Halifax County, were chief among those. These men, like the leaders of this cause elsewhere, made their contribution principally by selling the mulberry trees or by their promotional writings, rather than in the raising of worms and production of silk.[108] As a matter of

[103] A. L. Demaree, *The American Agricultural Press, 1819-1860,* 61-62.

[104] *Ibid.,* 63-65. Edmund Ruffin, perhaps more than any other editor, was inclined to talk down these recurrent crazes. *Ibid.,* 63-64. That some good resulted from these agricultural crazes is indicated in the following remarks by the editor of the *Prairie Farmer:* "Improvements in agriculture seldom come as *pure* improvements. They come in epidemics, mixed with a great deal of nonsense, and with more or less of humbug, and often of rascality." Quoted in *ibid.,* 65.

[105] *American Farmer,* IX (November, 1827), 261; *ibid.,* X (October, 1828), 229; *ibid.,* XII (April, 1830), 27.

[106] *Ibid.,* IX (November, 1827), 261.

[107] *Laws of N. C.* (1836-1837), 40-43.

[108] At the very height of this fever, the editor of the *Southern Cultivator* sought to undermine the enthusiasm for it by labelling it a humbug and by pointing out that ". . . mulberry trees, now-a-days, are grown for *sale,* not for feeding worms." Quoted in *Farmer's Advocate,* I (May, 1839), 273, with this editorial advice: "Reader, be cautious!! We think mulberry leaves would make rather unpalatable sallad without plenty of good bacon!!!" *Ibid.,* 282.

fact, the actual production of silk played an amazingly secondary role throughout the life of this fever.

Sidney Weller claimed, and to him should go, leadership in the state in promoting the *morus multicaulis* fever. He was the author of numerous articles on the subject, which were widely quoted, even by Edmund Ruffin.[109] Weller's interest in *morus multicaulis* was first attracted in the early 1830's by an article which appeared in the *American Farmer*. This led to the purchase, in 1833, for $1.00, of a plant which he set out on poor soil on his plantation. From this small beginning, he developed a nursery from which he sold $10,000 worth of stock by the end of 1838.[110] In 1839, Weller advertised his plants at $30.00 a thousand, "to be forwarded with due care to any place directed; at the risk of the buyer," for cash in advance.[111] Later, Weller started making a little silk himself and selling silk worm eggs. The collapse of interest in silk culture, he thought, resulted from the failure of experimenters to evolve an "American system" of culture.[112]

President Robert H. Morison, of Davidson College, in response to a request from the board of trustees, prepared an article setting forth the advantages the Piedmont section offered for the development of the silk industry. These were, as he observed, that very little land would be required; very little labor was used in the industry and the work was such that women and children could be more efficiently employed; large yield per acre; soil and climate were adapted to the culture of the mulberry; the demand for silk was constant; and, the industry offered opportunity for good profits.[113] President Morison also grew mulberry trees and offered them for sale "upon terms much lower than the Northern prices."[114]

Several planters capitalized upon the feverish demand for mulberry trees. Ebenezer Pettigrew and Josiah Collins, of Tyrrell County, made considerable profit by selling trees from a nursery they had.[115] J. Atkinson, of Pitt County, sold 20,000 young trees at forty cents each.

[109] *Farmer's Register,* VI (November, 1838), 492-493. A political opponent of Weller's, the editor of the Warrenton *Reporter,* sarcastically referred to his articles on agriculture, particularly on the mulberry tree, as "learned lucubrations," and said: "In fact his fame is as inseparately associated in the mind with the Morus Multicaulis Tree, as that of Gen. Jackson was with the Hickory." Quoted in Tarboro *Press,* May 22, 1847.

[110] *Farmer's Register,* VI (November, 1838), 492-493; Raleigh *Star,* October 31, 1838.

[111] *Farmer's Advocate,* I (March, 1839), 224.

[112] *Ibid.,* III (January, 1842), 229; *ibid.,* III (April, 1842), 318.

[113] Salisbury *Western Carolinian,* January 3, 1839. The editor of this paper was completely capitivated by Morison's arguments. He referred to it as follows: "It shows in a clear and forcible manner the great inducements held out for engaging in this branch of productive industry, which requires the investment of less capital, the expenditure of a smaller amount of labor, and renders a larger proportionate profit, than any other employment now engaged in." *Ibid.*

[114] Salisbury *Western Carolinian,* January 3, 1839.

[115] Tarboro *Press,* April 13, 1839.

Edmund Ruffin's comment on this sale was: "It is now a matter of congratulation, that this large parcel is purchased for planting in North Carolina."[116] As the profits continued good, the editor of the Raleigh *Register* became more enthusiastic. "Neighbors, try it!" he said. "Depend upon it, you will not regret the experiment."[117]

On the other hand, there is no evidence that anyone made any money out of silk culture in the state, or that more than mere "samples" of silk were produced. When this fever was at its height in 1839, Edmund Ruffin made a trip through the state. He saw no silk actually being made, but heard much talk of prospects in that direction.[118] These prospects never materialized, however. By 1842, this mania had run its course and died. One editor, in surveying the wreckage of this humbug, thought *morus multicaulis* might still be profitably cultivated for use in making paper since rags were scarce.[119]

Notwithstanding the total failure to gain anything of permanency from the *morus multicaulis* fever, the state's experience with it cannot be written off as a total loss. Coming as it did at a time when emigration from the state was at its height, and when the morale of the farmers was at its lowest ebb, it aroused an enthusiasm among those connected with it which was stimulating, and, for a few, it brought sizable profits.

In appraising the developments which occurred in the culture of fruits, vegetables, and other minor crops in North Carolina, it may be concluded that the state more than kept pace with the rest of the nation. In the absence of an effective commercial demand for these products, the aim of production in the state was overwhelmingly towards self-sufficiency. This purpose was substantially accomplished—there being little evidence of a scarcity of either fruits or vegetables.[120] There was improvement in the quality of these products as well as in the quantity. To a considerable extent, these results stemmed from the reforms instituted in agriculture. The state's position of leadership in the nation in the production of wine in the early 1840's gave North Carolinians something they could "point to with pride"—something that the morale of North Carolinians sorely needed at the time. The first honors won by North Carolina apples in national competitions, and the widespread popularity won by certain home-grown varieties of grapes also helped in the same manner. The work of individuals in developing and promoting these crops was outstanding. Men like Sidney Weller, Silas McDowell, Joseph Togno, Joshua Lindley, and S. W. Westbrook

[116] *Farmer's Register,* VI (October, 1838), 445.

[117] Raleigh *Register,* February 28, 1840.

[118] *Farmer's Register,* VIII (April, 1840), 254.

[119] Salisbury *Carolina Watchmen,* October 15, 1842.

[120] The per capita value of food crops produced in North Carolina in 1859-1860 was $53.00 dollars. This amount exceeded the average of the adjoining states and also the nation at large. A. O. Craven, "The Agricultural Reformers of the Ante-Bellum South," *American Historical Review,* XXXIII (January, 1928), 313 n.

were pioneers in the development of high quality fruits. Others, though less spectacular, also made contributions. To the mass of farmers, doubtless these accomplishments remained unknown. Few of them read agricultural journals, the annual reports of the Patent Office, or even the local papers. To them, the appearance of vegetable seed as an item of trade in the general store was taken as a matter of course. They bought little of these or the nursery developed trees and vines. For them, the tendency was to follow old practices and to oppose the new. Changes came but slowly.

CHAPTER IX

# CARE AND MAINTENANCE OF LIVESTOCK, 1820-1860

The practices employed in the raising of livestock in North Carolina proved more deep-seated and resistent to change than any other feature of the state's agriculture. In the colonial period, when population was scarce, grazing abundant, and little thought was given anywhere to the quality of animals, other than horses, the state had been a principal source of livestock, particularly cattle and swine, for the northern and West Indian markets. By 1820, however, this condition had changed. The state no longer was looked to as a source of meat supply. In fact, by that date North Carolina had become a buyer rather than a seller of such livestock as cattle, hogs, and draft animals. This position it continued to occupy, much to the mortification of the agricultural leaders and more intelligent farmers.

The reasons for this state of affairs are not difficult to find. Very few North Carolinians showed any interest in improving their livestock by selective breeding—a practice imported into the North from England, and, of which, by 1820 the benefits were being amply demonstrated in the high quality animals exhibited in livestock shows.[1] In the absence of any legal requirement that they be penned, North Carolina cattle generally continued to roam at will with only those selected for milking being kept near the owner. Breeding, of course, was haphazard and uncontrolled. Ownership of the offspring went to the owner of the mother. This was determined by brands and markings.[2]

With each expansion of the settled area of the state, there came a consequent reduction in the area of the "common" grazing land. Fewer and fewer cattle could be kept. Under the extensive system used everywhere in the colonial period and continued in North Carolina, the increase in value of a herd over and above the initial cost was almost clear profit. Even though profits ceased to flow, North Carolina stock raisers were very reluctant to change to a system that involved more expense and labor and offered not so high returns from the financial investment. The state ceased selling and started buying livestock. "We raise no stock of any kind except for home consumption," wrote a Lincoln County

[1] The Englishman, Robert Bakewell, initiated the practice of selective breeding of livestock. He was chiefly interested in increasing the quantity of beef. J. W. Thompson, *History of Livestock*, 3.

[2] R. G. Thwaites (ed.), *Travels West of the Alleganies Made . . . in 1802 by F. A. Michaux*, 299.

farmer, "and not half enough of that. . . ."[3] In Orange County the situation was not much better. "Neat cattle receive little attention here," it was observed, "cost little, and are worth little. What are called dairy cows sell at from $10 to $14 each . . . sheep very little attended to, and sadly annoyed and destroyed by dogs."[4]

Another factor that contributed to the failure to keep abreast of the times in the raising of livestock was the fact that no herdsmen were developed during the colonial period when the state led in the raising of stock. The men who handled the large herds of that period were mere drovers who gave no more attention to the cattle than was required by the periodic roundups, brandings, and drives to market places. When cattle-droving ceased paying them a competency they simply shifted to some other employment.

Undoubtedly, too, as planters became more staple-conscious they tended to be less thoughtful of the other interests of their plantations. Many of the best planters even failed to make any attempt to raise their own work stock. Stephen A. Norfleet, for example, leading planter in Bertie County and to whom frequent reference has been made in this volume, paid out $2,012 between February, 1859, and March, 1860, for thirteen young mules and two horse colts.[5]

The interest of some planters, too, was so concentrated elsewhere that they failed to raise enough provender for their livestock. John S. Dancy, of Edgecombe County, owner of one of the best cotton plantations in the state, first president of the state agricultural society, and in most matters an "improving" planter, shame-facedly admitted that he was one of that group.[6]

Ultra-conservatism and just plain laziness played a part also. These characteristics are reflected in the following remarks of Calvin Jones, of Wake Forest, to a query as to the prospects of sheep raising becoming an important feature of North Carolina agriculture.

> There can be no doubt that keeping sheep on a large scale, would be a very profitable business; but the experiment must be made by persons from abroad; for as our lands and negroes afford to most a competency, without the pain of exertion, or the hazard of enterprise, few motives can be found powerful enough to overcome our habits, or change our pursuits.[7]

Finally, through either mis-management or over-management high quality livestock imported into the state, with the idea of improving the local stock, quite often failed to come up to expectations as to appearance

[3] *Report of the Commissioner of Patents* (1851), 315.

[4] *Ibid.* (1850), 376.

[5] Norfleet Diary, *passim,* Norfleet Papers.

[6] "I confess (with sorrow be it said) to using Hay from the North one season and I assert that *well cured* crab grass Hay is in every respect equal and often superior to the yankee article." Tarboro *Southerner,* October 7, 1854.

[7] *American Farmer,* X (November, 1828), 266.

and performance, or died soon after arrival. This stock was usually bought at fairs in the North. At the time of purchase, they were in a fattened condition for the purpose of making a good show. The purchaser, in an effort to keep the animal in that condition, often overfed it with the result that the animal died.[8] The actual management of this stock often devolved upon slaves who were given the task not because of any special fitness, but because they were unable through either age, sickness, or deformity to do a field hands work. The quality of the food given this stock was often coarse and led to degeneracy in appearance and performance. As a recent student has said, in the raising of improved livestock the ante-bellum North Carolina farmer was just emerging from the frontier stage, was unaccustomed to keeping good pastures, and supplying sufficient shelter, and frequently failed because he did not appreciate the fact that "improved cattle called for improved attention."[9] Some of the improved breeds of cattle did not thrive in the climate.[10] Each failure, for whatever cause, served to discourage others from trying to improve their stock by importations of high quality animals. In the breeding of horses alone, North Carolinians showed a determination to have the best blood. For all other kinds of livestock, the efforts at improvement by this means were scanty and usually not determined enough.

Agricultural reformers and the friends of improved agriculture sought to correct these conditions by every means known to them. Ridicule and sarcasm were heaped upon the evils of the system being used. Appeals were made to the intelligence, the sectional partiotism, the pocketbook, and the personal pride of stock raisers. Good examples were pointed out and emulation encouraged. Stock shows were held and prizes offered for high quality animals. The agricultural press was filled with advice on the proper management of livestock and the benefits that would flow therefrom.

An effort was made to arouse an enlightened self-interest in this respect. "Pay less attention to the breeding of hounds," wrote a correspondent of the Rutherfordton *Enquirer,* "and more to that of horses. The sleek fat beeves and golden rosy flavored butter of New York, are not the produce of broom grass and scrub pines."[11] Charles Fisher, in a speech before the Rowan County Agricultural Society, said: "Is it not a fact, highly discreditable to the large, populous, and wealthy county of Rowan, that it cannot supply itself with beef? For the past 8 or 10 years, not fewer than between 2 and 300 beeves, in each season, have

[8] *Farmer's Journal,* I (April, 1852), 23; *North Carolina Planter,* I (April, 1858), 118-119.

[9] Charles T. Leavitt, "Attempts to Improve Cattle Breeds in the United States, 1790-1860," *Agricultural History,* VII (April, 1933), 60, cited hereinafter as "Attempts to Improve Cattle Breeds."

[10] *Ibid.,* 57, 60.

[11] Reprint in *North Carolina Planter,* I (March, 1858), 83-84.

been brought from the upper counties and slaughtered for the market of this small town."[12] "Without the hazard of refutation, I may say," he continued, "that there is not at this day, a first rate milch cow in the county of Rowan; nor will there be, until an entire change takes place in our system of management . . . *One Cow,* well taken care of, will supply more milk than *three* badly kept. . . ."[13] One editor drew attention to an article he had printed on the management of a dairy herd in Scotland and urged his readers to "contrast it with the management of cows in . . . the southern states—where, as we once heard it remarked, if, after milking six cows, the bottom of the pail should be *wet,* the milkmaid thinks it sufficient encouragement to go on and milk the rest!!"[14] "Never keep a poor or malformed animal of any kind," advised another editor. "It is better to kill or give away such, than incur the expense of keeping them, and the risk of having their peculiar deformities communicated to the rest of your stock."[15] In an article entitled, "Obligation to Brutes," an editor wrote:

> We may use them for labor, or for innocent physical recreation, as when we employ the horse for draught or for the saddle. But while we so use them, we are bound to treat them kindly, to furnish them sufficient food and with convenient shelter. He who cannot feed a brute well ought not to own one.[16]

The following are examples of "Fixed Facts in Agriculture" which were widely published in this connection:

> The chopping or grinding of grain, to be fed to stock, operates as a saving of at least twenty-five per cent.
>
> By stabling and shedding stock through the winter, a saving of one-fourth of the food may be effected—that is, one-fourth less food will answer, than when such stock may be exposed to the inclemencies of the weather.
>
> Milk cows, in winter, should be kept in dry, moderately warm, but well ventilated, quarters, be regularly fed and watered three times a day, salted twice or thrice a week, have clean beds, be curried daily, and, in addition to their long provender, should receive succulent food, morning and evening.[17]

Similar bits of advice and criticism were constantly given covering every aspect of the livestock industry. Nevertheless, it is rather certain that farmers generally were more persistently neglectful of improved methods in this interest than in any other pertaining to their operations.[18] From 1820 to about 1840, stock raising in North Carolina continued in decline. In fact, instead of being a source of revenue to the state it was

[12] *American Farmer,* III (August, 1821), 170.
[13] *Ibid.*
[14] *American Farmer,* II (February, 1821), 373.
[15] Hillsboro *Recorder,* November 8, 1848.
[16] *Ibid.,* August 22, 1849.
[17] Reprint from *American Farmer* in *Farmer's Journal,* I (August, 1852), 131-132.
[18] *Farmer's Journal,* I (October, 1852), 208.

a source of financial loss. It was not until after 1840 that improvement in this respect became perceptible.[19]

A recent student of the problem has said that before there can be any considerable improvement in the breeds of livestock, two fundamental factors are necessary: first, sufficiently high profits resulting from good markets or speculative eras; and, secondly, the development of so-called improved breeds.[20] Neither of these conditions prevailed in ante-bellum North Carolina. When livestock raising flourished in the state, the proprietors operated the business in the same manner as the farmer who tilled his land continuously in a crop until it ceased to pay for the necessary labor and then abandoned it. They put nothing back into the business. With them, it was a purely extractive enterprise. They also did nothing to improve the breeds or quality of their products. As a result of their failure to do anything to promote the permanency of the industry, decline became inevitable.

For the nation at large, note has already been taken of the introduction of mules and the importations of highly blooded horses and Merino sheep. The first breed of cattle imported for the purpose of improving the American stock was the Shorthorn, which was brought from England into Virginia in 1783. Their value was immediately recognized and numerous additional importations were made.[21] This breed was followed by Durhams, Devons, Ayrshires, and Herefords. Some of all these breeds soon found their way into North Carolina. At the first livestock show held in the state at Salisbury, in 1821, the first three prizes went to five head of Devon cattle which belonged to a farmer of Rowan County.[22]

Among those who worked for improvement in the quality of livestock in the state, two schools of thought developed. There was one group which labored for a complete revolution by advocating a complete change of breed.[23] George W. Jeffreys, who was one of this group, pointed out that different breeds were adapted to different purposes and that farmers should be guided by that fact in making their selections.

> A breed of cattle equally well adapted to the butcher, to the dairy, and to the plow or cart is nowhere to be met with . . .these properties are hardly consistent with each other, and belong to animals of different forms and proportions. It should be the object of farmers to have all their stock of

[19] *Compendium . . . From the Returns of the Sixth Census. . . .*, 178-180; *The Seventh Census of the United States: 1850. . . .*, 319-320; *U. S. Census of 1860* (Agriculture), 108-109. See also, J. W. Thompson, *History of Livestock,* 140.

[20] C. T. Leavitt, "Attempts to Improve Cattle Breeds," *loc. cit.,* 51.

[21] C. T. Leavitt, "Attempts to Improve Cattle Breeds," *loc. cit.,* 58; George M. Rommel, "The Animal Industry of the South," *The South in the Building of the Nation,* V, 246.

[22] W. H. Gehrke, "The Ante-Bellum Agriculture of the Germans in North Carolina," *Agricultural History,* IX (July, 1935), 150.

[23] *American Farmer,* III (August, 1821), 170.

animals of the best breeds: and to study useful qualities, more than shewy figures.[24]

Throughout the period there were a few who followed this advice and sought to build new herds by inter-breeding only the highly blooded stock which they had purchased from out of the state.

On the other hand, the high fatality rate among the imports of blooded stock caused such heavy financial loss and discouragement that others recommended a more gradual approach towards improvement. Paul C. Cameron, of Orange County, thought that no attempt should be made to improve the quality of livestock "by costly importations," until the individual farmer was prepared to provide full supplies of proper food and "a more humane care." Then, he would effect the improvement gradually by cross-breeding the blooded animals with "judicious selections" from the native stock, "which embraces a few of the best, and many of the worst, and almost every variety of shape, color and character."[25]

It was recognized by some that the disadvantages attending the development of blooded herds were increased tremendously by the lack of knowledge as to how such animals should be managed. For this reason, they chose to improve their herds by breeding the blooded stock with the best native animals whose hardihood was known. ". . . it is but a species of monomania," wrote a Wake County correspondent, "for any man to attempt to change the entire stock of any country or even district of country at once. The change must be gradual and keep pace with the knowledge of the nature, habits and character of the newly introduced breeds."[26]

The mass of farmers, however, evinced but little interest in the improved breeds and persisted in managing their livestock in the same manner as did their pioneer forefathers. Under their system, the stock were allowed to run wild with a minimum of provision being made for their feeding, shelter, health, and breeding. Winter time worked tremendous suffering upon stock neglected in that manner. Disease, starvation, and wild animals took a heavy toll in lives.[27] A Warren County farmer, who handled his stock in this manner, admitted to Edmund Ruffin, in 1837, that "a large proportion on most farms died" during the winter, and he himself "calculated on losing five or six, on an average, out of his small number of stock every winter."[28]

Milk cattle suffered particularly from such lack of care and attention. The following description of those near Raleigh, drawn in 1856, is

[24] (George W. Jeffreys), *Essays on Agriculture,* 203.
[25] *Arator,* I (June, 1855), 71.
[26] *North Carolina Farmer,* I (December, 1845), 147.
[27] Adelaide L. Fries (ed.), *Records of the Moravians,* VI, 2824-2825; *Farmer's Register,* VII (November, 1839), 701; *ibid.,* V (August, 1837), 230.
[28] *Farmer's Register,* V (August, 1837), 230.

thought to be typical of conditions commonly found throughout the state.

On almost every road leading into the City, one can see, any evening, long files of lean, gaunt, half-starved cattle, returning from picking up a very scanty portion of forage in the woods, to yield an equally scanty supply of a watery fluid, nicknamed 'milk,' for their owner's supper. . . . Compared with a thorough bred, they do not seem to belong to the same species. . . .[29]

Some farm families, milking as many as six or more of such ill-fed animals, sometimes had to buy butter for their tables.[30]

As to the percentage of farmers who demonstrated no interest in improving their cattle, the estimates range from nine-tenths to nineteen-twentieths.[31] With regard to the breeding of milk cattle, according to a contemporary critic, this group paid "no attention whatever to their fatting or milking qualities—make no effort, by careful selections, liberal feeding, kind and generous treatment, to bring these valuable animals to any state of perfection."[32] In this group, too, were some who evinced open hostility to the improved breeds and labeled the idea that they were superior to the common stock a humbug.[33]

A correspondent from Weldon, who described himself as being in the category of "pinhook farmers," illustrates very well the reaction of the mass of farmers in the state to the movement for the improvement of milk cattle and agriculture generally. He admitted being "bothered" by the import of improved cattle, hogs, and sheep into the state—"things that very few of us are able to buy and ought not if we were able; first, because they are too high; and secondly, because our own cattle will answer very well if we take proper care of them. . . ."[34]

With this class of farmers, the whole cause of improvement of livestock, as well as of agriculture generally, suffered from the results of overly enthusiastic and too extravagant promotion by the reformers. With each failure of a breed of cattle or of a new farm implement to come up to expectations they rejoiced for they were opposed to change in the first place. This group, too, claimed that positive injury was done them by promoters who stimulated a too great confidence in innovations which later failed. According to one correspondent:

It . . . causes us to overlook many of the advantages around us, and if we do not entirely lose sight of the conveniences at our own homes, we lose confidence in them, because they are too common—are not far enough fetched and have not made a great flutter in the world.[35]

[29] *Arator,* I (February, 1856), 338.
[30] Reprint from Louisburg *Eagle* in *Arator,* I (December, 1855), 277.
[31] *Farmer's Journal,* I (October, 1852), 208-209; *North Carolina Farmer,* I (February, 1846), 195.
[32] *North Carolina Farmer,* I (February, 1846), 195.
[33] Reprint from Louisburg *Eagle* in *Arator,* I (December, 1855), 277.
[34] *North Carolina Planter,* II (January, 1859), 3.
[35] *North Carolina Planter,* II (January, 1859), 3.

This was undoubtedly true, as agricultural reformers must have recognized, for, in the latter part of the period, the reform and promotional literature carried less extravagant claim, was more cautious, and restrained.

In livestock improvement, too, the number of those who would approach the goal gradually, increased. The following remarks by the editor of the *North Carolina Farmer* are illustrative of the views of that group.

To substitute the improved English breeds for our own entirely, is neither practicable nor desirable. The most that can be expected, or that would be proper to be attempted, is to improve the native stocks already among us, and to cross them highly and judiciously with the best foreign bloods. This plan, pursued with patience and perseverance, will enable the poor or moderate farmer, as well as the rich, to secure the benefits of a highly improved stock of cattle.[36]

As agencies in effecting reform in this manner, it was proposed that companies of interested farmers be organized or that agricultural societies undertake the importation of blooded stock for interbreeding with the common breeds.[37]

Doctor William R. Holt, proprietor of Linwood Farm in Davidson County, was, perhaps, outstanding in the amount of contribution made towards livestock improvement in the state.[38] His imports of blooded livestock included cattle, horses, and sheep. Fine pastures were provided for their grazing in summer and the best of home-grown hay for their winter feeding. Of cattle, he had thoroughbred Devons and Durhams. There were over a hundred of the former in his herd in 1853, all of which looked "finely."[39] His Durhams were described as "of unsurpassed beauty," and attracted considerable interest wherever exhibited. A local editor thought it remarkable "that out of such a large number, there should be no difference in the color of the skin and hair, and no scrubs among them. All were of red hair and yellow skins, clean limbs, and every way handsome."[40] Doctor Holt exhibited some of his cattle at the state fair, where they won most of the premiums, and were

[36] *North Carolina Farmer,* I (June, 1845), 22.

[37] *Ibid.*

[38] Described by one editor as "a really scientific, practical farmer, well informed upon every subject in connection with agricultural improvement," Doctor Holt, by using the best implements and the most advanced methods, improved his previously worn out farm until it became one of the best in the state. *Farmer's Journal,* II (September, 1853), 177. Doctor Holt was actively interested in general agricultural improvement, being one of the leading promoters of and livestock exhibitors at the state fair, and president of the state agricultural socity in 1859. Many visitors, including George Bancroft, the historian, were attracted to Doctor Holt's farm. W. H. Gehrke, "The Ante-Bellum Agriculture of the Germans in North Carolina," *Agricultural History,* IX (July, 1935), 160.

[39] *Farmer's Journal,* II (September, 1853), 177.

[40] Salisbury *Carolina Watchman,* May 9, 1850.

eagerly bought at prices considerably above the market prices of other cattle by people interested in improving their herds.[41] Doctor Holt was unremitting in his efforts to improve livestock in the state and won for himself first rank as a public benefactor.[42]

As far as numbers are concerned, the state ranked twelfth in the Union in population in 1860, and thirteenth in the number of milk cows and other cattle. The latter position represented a decline in rank during the decade of the 1850's. The poor quality of the milk cows was reflected in the fact that the state ranked only twenty-fourth, in 1860, in the production of butter and cheese.[43] As repeatedly pointed out by reformers, the average North Carolina farmer had too many cows to give them the care and attention which they should have had.[44]

Although raising of the ordinary breed of sheep was carried on throughout the state, and certain individuals participated in the craze for the Merino breed, which swept over the nation in the first two decades of the century, sheep raising did not develop into a large-scale business in ante-bellum North Carolina.

Several factors help to explain the lack of growth of this industry, particularly in the Mountain section which was thought by some to be very well adapted to its development. In the first place the price of wool, which had risen to tremendous heights as a result of the Embargo and the failure of European sources of supply resulting from the Napoleonic Wars, fell precipitately when peace came.[45] With the collapse of wool prices, the popularity of Merinos, which in 1810 had sold for as high as $1,000 each and whose wool had brought $2.00 a pound at the same time that common wool brought only forty cents, entered a period of sharp decline.[46] In the North, it has been said, "farmers turned against the Merinos with a disgust as unreasoning as had been their former enthusiasm."[47] Before the end of the decade of the 1820's, full-blooded Merinos were selling in that section for less than $1.50 each.[48] This development discouraged sheep raising everywhere.

Based upon the animal's lack of hardiness to thrive under the sort of neglect shown other livestock, a prejudice against the sheep had become fixed in the population. It was thought to be a too delicate animal. The early agricultural reformers, in their remarks upon sheep-raising, actually fostered that prejudice. John Taylor, who had experimented with

[41] *Farmer's Journal,* II (November, 1853), 240-241.

[42] Salisbury *Carolina Watchman,* May 9, 1850.

[43] *U. S. Census of 1860* (Agriculture), 105-111.

[44] *Farmer's Advocate,* III (January, 1842), 230; *Farmer's Journal,* I (October, 1852), 208-209.

[45] Percy Wells Bidwell and John I. Falconer, *History of Agriculture in the Northern United States, 1620-1860,* 219.

[46] *Ibid.,* 217-219.

[47] P. W. Bidwell and J. I. Falconer, *History of Agriculture in the Northern United States, 1620-1860,* 219.

[48] Tarboro *Press,* September 25, 1829.

sheep raising on a rather large scale for sixteen years, concluded that sheep ate more for their size than any kind of livestock, were more liable to death by disease or accident, and that they could not be made an object of profitable culture in those regions which had a warm, dry climate and sandy soils. Their chief worth, he thought, lay in the high quality of their manure.[49] George W. Jeffreys echoed those sentiments and added the inference that, except for the meat they afforded, sheep would be of little value in any region "where cotton can be raised so easily."[50] John Steele, influential planter of Rowan County, when asked by a correspondent for advice on sheep raising, warned him against "such frivilous speculations." The following remarks from Steele's reply, which he said were based upon his own experience with the animal, illustrate the depth of feeling against sheep raising.

> Of all agricultural stock I have found sheep to be the most precarious and unprofitable. Keep them fat they become disordered and die. Keep them lean, it is worse. House them in the winter, they languish, leave them unsheltered the dogs destroy them. Confine them in fields of rich pasture during the same they become disordered,—turn them without your fences they stray away, and in their best condition with good management . . . the profit is very insignificant when compared with other objects of our attention.[51]

Undoubtedly the greatest discouragement to sheep raising grew out of the ravages made against the pack by dogs and wild animals, particularly dogs, of which, the number was so great as to be a public nuisance.[52] Very few people owned large enough flocks to be able to provide full-time shepherds which made it an easy matter for vicious dogs to annoy and kill the sheep.[53]

Numerous efforts were made to secure legal protection against these losses, and those suffered from wild animal pests, but little was accomplished. In 1817, commissioners of incorporated towns were authorized to impose tax upon dogs kept in the towns.[54] Several counties secured authorization to pay out from county funds certain bounties for panthers and wolves killed within their limits. The primary purpose of

[49] From reprint of *Arator* in *Farmer's Register*, VIII (December, 1840), 743. Charles Fisher alleged that the average farmer was more neglectful of his sheep than any other branch of livestock; and, as a result, the average production of wool per sheep in North Carolina was only one and one-half to two pounds, whereas, in the North, production ranged from six to twelve pounds. *American Farmer*, III (August, 1821), 170.

[50] (George W. Jeffreys), *Essays on Agriculture*, 207-208.

[51] H. M. Wagstaff (ed.), *Papers of John Steele*, II, 629.

[52] *Southern Agriculturist*, II (May, 1829), 223-224; *Farmer's Journal*, II (January, 1854), 319; *ibid.*, II (February, 1854), 340.

[53] A traveler through the Piedmont section in 1802 observed that a person owning as many as a dozen sheep was considered to have a large flock. R. G. Thwaites (ed.), *Travels West of the Alleganies Made . . . in 1802 by F. A. Michaux*, 300.

[54] *Laws of N. C.* (1817), 25.

these acts was to encourage sheep raising.[55] Agricultural societies became interested in curbing this menace to sheep raising. In 1854, the state society adopted a memorial and designated a committee to present it to the legislature, in which, it was proposed that a state wide tax be imposed upon dogs as "a matter of the most vital importance to the advancement of sheep husbandry in our state. . . ."[56] In 1855, the legislature passed an act which made the dog owner or slaveowner, if the dog belonged to a slave, responsible up to an amount twice as great as the real damage his dog caused.[57] Had such legislation been passed earlier, the whole aspect of sheep raising in the state might have been changed.

Notwithstanding the failure to become a large scale commercial interest, sheep raising was widely carried on in the state. Nearly every family raised its own supply of mutton and wool. The wool was chiefly mixed with cotton in the manufacture of linsey-woolsey, from which a great part of the winter clothing of slaves and whites was made.[58] Although the Piedmont counties led in production in 1860, only four of the state's eighty-six counties had fewer than 2,000 sheep each. Between 1840 and 1860 the number of sheep in the state remained about the same.[59] In relative rank in the Union in sheep production, however, the state fell from twelfth to fifteenth place between 1850 and 1860.[60]

The prospects for establishing a woolen industry were discussed in the state.[61] Developments in that direction came very slowly, however. In 1860, there were seven establishments in the state using wool in the making of cloth. Over one-half million pounds of wool were used in that year, from which, 639,000 yards of cloth were made. In volume, this business increased slightly over 300 per cent in the decade of the fifties.[62] This development, however, came too late to exercise an appreciable influence on sheep raising. In North Carolina, as in the other southern states, although inquiry on the subject was manifested throughout the period, sheep husbandry did not take hold and prosper.[63]

Being less badly affected by the neglect with which most farmers treated their livestock, swine proved more satisfactory and beneficial to the average ante-bellum North Carolinian than any other form of livestock. "No domestic animal contributes so largely to our support,"

[55] *Ibid.* (1828-1829), 86; *ibid.* (1834), 92; *ibid.* (1835), 115.
[56] *Farmer's Journal,* III (November, 1854), 243.
[57] *Laws of N. C.* (1854-1855), 115.
[58] *Niles' Weekly Register,* XLIII (February 16, 1833), 423-424; *American Farmer,* X (November, 1828), 265.
[59] *U. S. Census of 1860* (Agriculture), 105-111.
[60] *Ibid.,* xxix.
[61] See reprint from Raleigh *Register* in *American Farmer,* IX (December, 1827), 307.
[62] *U. S. Census of 1860* (Manufactures), xxxv.
[63] J. D. B. DeBow, *Industrial Resources,* III, 188-195; J. W. Thompson, *History of Livestock,* 134.

wrote George W. Jeffreys, "and that of the laborers employed on our farms, as the hog; and it is fortunate that none multiplies so fast, grows so rapidly, is as thrifty, hardy, and as easily raised."[64] Having these qualifications, the hog fitted very well into the pioneer's way of doing things, and great numbers of them were raised at little expense other than the corn that some farmers threw out to them at fattening time.

As the area of settlement in the state increased, the forests, with their supplies of mast upon which swine foraged, decreased. Fewer hogs could be raised by following the methods previously employed, and more supplemental feeding had to be provided. As a general rule, hogs were not slaughtered until they were two or three years old. During most of their lives they were allowed to run wild and seek their own food.

Those selected for slaughter were penned up one or more months for fattening and fed principally on corn. This was usually done in a slovenly and wasteful manner too. The pens were made of fence rails, enclosing a stream of water if possible, and into this the fattening hogs were put. Corn was thrown to the hogs on the cob and usually a large part of it was lost in the quagmire which they made in the pens.[65]

This method of penning and feeding the fattening hogs was criticized by the better farmers as "a pernicious practice" which not only caused waste, but injured the hogs. A "close pen" was recommended in its place. This system possessed several advantages. Feeding was less wastefully done. Less feed was required since the animal's exercise was considerably reduced. Also, the manure was not lost.[66]

Much more attention was also given to the hog's diet. One farmer chided the editor of the *Farmer's Journal* for not giving more information on the subject since it was "an all-absorbing topic of the farmers of the country."[67] Another wrote of "edging in a little book learning occasionally upon the subject," being chiefly a "practical or experimental" farmer himself.[68] This man fattened his hogs by feeding them on peas for a week, and then changing over for a week to sweet potatoes, and then back to peas with "an occasional bait of corn" and waste from his kitchen. When the peas and potatoes became scarce, the hogs were close-penned and fed on boiled feed, chiefly corn, peas, and potatoes.[69] The practice of cooking food for swine was introduced and recommended as of great benefit in shortening the fattening period.[70] Pea-

[64] (George W. Jeffreys), *Essays on Agriculture,* 205. John Taylor's opinion of the hog was almost identical with the above. See reprint of *Arator* in *Farmer's Register,* VIII (December, 1840), 745.

[65] *American Farmer,* III (August, 1821), 170.

[66] *Farmer's Journal,* II (January, 1854), 305-306.

[67] *Farmer's Journal,* II (January, 1854), 308.

[68] *Ibid.,* II (March, 1854), 371-372.

[69] *Ibid.*

[70] *Ibid.,* II (January, 1854), 305-306.

nuts, wheat "shorts" or bran, pumpkins, fruits, and clover were also used as hog feed.[71]

The transition from the extensive practices, involving very little feeding of swine which prevailed in colonial North Carolina, to the intensive feeding of the animal required by changed conditions in the ante-bellum period, was not easily made, and, for some years in the latter period, the state was not self-sufficient in pork.[72] Some farmers even had to buy pork. This situation was severely criticized by the agricultural reformers and regarded with shame by the more provident farmers. According to one of the latter group:

> One reason for this is, that no provision is made for raising swine, beyond the cultivation of a few extra corn field peas and sweet potatoes, save the scanty allowance of corn which is occasionally thrown to them on the ground; and many for 9 months in the year don't even get this: with them it is 'root pig or die.' This is absolutely scandalous—it is unmerciful and sinful.[73]

In an effort to improve conditions, emphasis was given to the ideas that instead of keeping a hog two or three years, "half perished all of the time," it should never be kept through more than one winter; it should be intensively fed; and, only the most "thrifty" should be selected for breeding.[74]

For those who adopted this system, the benefits soon became apparent. A group of farmers in Currituck County formed an association which offered cash premiums to the members producing the best hogs in several age brackets. At the expiration of the contest time, twenty-five hogs were exhibited. Among the winners, the one-year-olds clearly outdistanced all others in the amount of weight put on for their age.[75] It was found that the common stock of hogs, if given proper attention as to penning, feeding, and selection in breeding, produced a much better formed and more valuable animal.[76]

Some attention, too, was given to securing the improved breeds of

[71] *North Carolina Planter,* I (March, 1858), 69-70; (George W. Jeffreys), *Essays on Agriculture,* 205-207; *Report of the Commissioner of Patents* (1849), 143; Norfleet Diary, *passim,* Norfleet Papers.

[72] Most of the imports of pork came from Tennessee and Kentucky—the hogs being driven, on foot, through the western part of the state. Much the larger number of these animals went to South Carolina and Georgia, but large numbers also found a market in the state. See *Carolina Planter,* I (February, 1840), 50; *DeBow's Review,* XVI (May, 1854), 540; *North Carolina Farmer,* II (August, 1846), 50-52.

[73] *North Carolina Farmer,* II (August, 1846), 50.

[74] *Farmer's Advocate,* III (January, 1842), 230.

[75] *American Farmer,* IX (January, 1828), 355. The first three winners among the one-year-olds averaged 325 pounds net. Of course, these pigs had been given far above average feeding. The average weight of North Carolina pigs in this age bracket was estimated at 130 pounds. *Report of the Commissioner of Patents* (1849), 143.

[76] *Farmer's Journal,* I (March, 1853), 379; *ibid.,* II (February, 1854), 346.

hogs. The Chinese hog, which attained some popularity in the 1820's, was introduced into the state.[77] There is no evidence, however, that its popularity became widespread. The Berkshire, Russian and Kenilworth hogs were also introduced. The Berkshire proved most popular and was widely cross-bred with native and other imported hogs.[78] The "No-bone" breed was introduced at the end of the period.[79]

In the proposals concerning the raising of swine, there were no complaints raised, even by the reformers, as to the quality of the bacon, hams and other pork products whether produced from the common or improved breeds. In fact, when mentioned at all, the quality of the meat was generally highly praised.[80]

All indications are that tremendous quantities of pork were consumed. One planter estimated that the annual consumption of pork per person in the state equalled two hundred pounds.[81] A traveler through the state observed that pork was always served and sometimes prepared in as many as seven different ways for one meal.[82]

In spite of the heavy consumption, the state had unquestionably regained a position of self-sufficiency in the production of pork by 1860. Imports from Kentucky and Tennessee had been reduced to a mere trickle of what they had been formerly. Besides, many North Carolinians were shipping pork to purchasers outside the state. In 1859, only five of the state's eighty-six counties tallied fewer than 10,000 swine. In the entire state, at that date, there were about two and one-half hogs per person including both blacks and whites.[83] In relative rank in the Union, the state advanced during the fifties from ninth to eighth place.[84] Of greater importance than mere numbers, considerable improvement had been accomplished in feeding and breeding practices which increased the value of the individual hog.

Notwithstanding the progress made, many swine continued to live in a semi-wild state, receiving no attention as to breeding, feeding, or shelter, and often unidentified as to ownership. The number of these is indicated by the fact that after the formal count for the 1860 census had been made, it was estimated that 206,976 swine should be added to the number already obtained.[85]

[77] *American Farmer,* III (August, 1821), 170.

[78] Salisbury *Carolina Watchman,* January 7, 1843; *ibid.,* February 4, 1843; *Report of the Commissioner of Patents* (1850), 376.

[79] *North Carolina Planter,* I (April, 1858), 118.

[80] Charles Fisher, for example, considered the methods used in the raising of swine wasteful and "wretched" in most respects, but the pork to be excellent. *American Farmer,* III (August, 1821), 170. The task of curing the meat after slaughter, even on the well-to-do plantations, usually was performed or supervised by the house-wife.

[81] *Report of the Commissioner of Patents* (1849), 143.

[82] F. L. Olmsted, *Journey in the Seaboard Slave States,* 326-327.

[83] *U. S. Census of 1860* (Agriculture), 105-111.

[84] *U. S. Census of 1860* (Agriculture), xxix ff.

[85] *Ibid.,* 192.

In regard to draft animals, two significant developments occurred: the gradual replacement of the horse for work purposes on the farm by the mule and ox; and an increasing agitation for self-sufficiency in the production of work animals.

The mule proved particularly well adapted to perform the work required of a draft animal, and to withstand the neglect usually accorded livestock on the ante-bellum North Carolina farm. As a result, that animal gradually grew in favor and was produced or purchased in large numbers. It naturally competed with the horse and many articles appeared in the press comparing the qualities of the two.

Most of those articles favored the mule. The following remarks by Calvin Jones are typical of the comments upon that animal.

> They are easily broke to labour, and if kindly treated, are very docile; but they are sensible of ill usage, and apt to requite it. They can endure heat and fatigue better than horses; are more true to the draft, and less liable to diseases. They live also to a greater age; cost less in the raising; require only two thirds the food when at labour; and when turned out to range, can live and fatten where horses would starve and die.[86]

Paul C. Cameron wrote: "It is but naked justice for me to declare, that for all the drudgery of the farm, the Mule is without an equal . . . No animal is more worthy of our care. . . ."[87]

The advocates of the ox listed about the same advantages over the horse as were used for the mule; and besides, the ox, after its days of usefulness as a draft animal were ended, was usually fattened and eaten. George W. Jeffreys, however, noted "a strong prejudice against this generous animal" which considerably retarded its rise in popularity.[88] Although definitely more economical for general farm purposes than the horse, the ox fell far short in speed of movement.[89]

From 1850 to 1860, the ox increased only slightly faster than the horse, whose number remained almost stationary; whereas, the mule increased over 100 per cent. In spite of this increase, however, there were still three times as many horses in the state in 1860 as there were either mules or oxen.[90] In fact there was an average of only a little more than one-half a mule and one-half an ox per farmer in the state.[91] Undoubtedly, the horse led all others in the performance of draft func-

[86] *American Farmer,* X (November, 1828), 273.

[87] *Arator,* I (June, 1855), 73. For similar remarks see *Farmer's Journal,* I (July, 1852), 123-124; *ibid.,* III (September, 1854), 186.

[88] (George W. Jeffreys), *Essays on Agriculture,* 198-200.

[89] See reprint on this subject from *Skinner's Cattle Doctor* in *North Carolina Farmer,* III (May, 1848), 265-269.

[90] There were 150,661 horses in the state in 1860. *U. S. Census of 1860* (Agriculture), 108.

[91] There were 85,198 farmers and 121 planters listed in the state in 1860, and only 51,388 asses and mules, and 48,511 oxen. *Ibid.* (Population), 362-363.

tions on North Carolina farms, and there were many who still considered it superior to the mule and ox for that purpose.[92]

Of these three animals, the state produced only enough oxen to meet its own requirements. The purchases of horses and mules constituted a constant drain of capital away from the state which might otherwise have been used in the improvement of its agriculture.[93]

This state of affairs was severely criticized by the agricultural reformers. Doctor John F. Tompkins, who said that in this case the farmers of the state had "been guilty of improper conduct long enough," wrote as follows:

> It seems to us very strange, that while the farmers of every other state in the union are turning their attention to the raising of their own horses and mules, those in North Carolina still continue to be dependent upon other states for their supply of this kind of stock. We have frequently called the attention of individuals to this matter, and they say, as a general thing, that it will not pay; but when we ask them if they have ever tried it, they answer in the negative.—We contend that by proper management it will pay; there is nothing either in the climate or products of our state to prevent it.[94]

Paul C. Cameron said the stock of horses in the state were "all wofully deficient in all the essentials of size, figure, bone, muscle, depth of girth, game and action," a condition which resulted from the fact that all previous efforts in horse-breeding had been "directed by a single idea, and controlled by a single object—the attainment of the highest speed."[95] What the state needed, he said, was a work horse, not a racing animal. He urged farmers to raise their own mules, believing like some others, that those raised locally were superior to the imported ones.[96]

A lively interest was shown in the breeding of race horses, but it was only in the latter years that attention was given to the production of "improved" work horses. The Vermont Morgan and the Cleveland Bay breeds of work horses were particularly recommended.[97] The former was described as "celebrated for all practical purposes, (except

[92] "Anti-Mule" complained in the *Southern Planter* against the tendency to replace horses with mules, "which cannot keep up their own race," and cited figures in an effort to show that the horse was more economical. Reprinted in *Farmer's Journal,* II (March, 1854), 356.

[93] Stephen A. Norfleet, of Bertie County, whose plantation in nearly all respects was self-sufficient, spent sometimes more than ten per cent of his annual gross income for mules and horses. For the mules, which he used for draft purposes, he paid an average of $150 each during the fifties. The horses were used principally for saddle and coach purposes and averaged costing about $200 each. Norfleet Diary, *passim,* Norfleet Papers.

[94] *Farmer's Journal,* II (June, 1853), 83.

[95] *Arator,* I (June, 1855), 72.

[96] *Arator,* I (June, 1855), 73. Calvin Jones said the North Carolina raised mule was "hardier" than those coming from a colder climate. *American Farmer,* X (November, 1828), 273.

[97] *Arator,* I (June, 1855), 72; *North Carolina Planter,* I (April, 1858), 117.

a running horse,) for beauty, style, action, draft and speed—very healthy and hardy—and with an eye that no other horse has."[98] Several of the Morgan stallions were sold in the state during the fifties and proved highly satisfactory.[99] "Goldfinder," a pure bred stallion of the Morgan stock, attracted considerable interest when exhibited in Raleigh in 1858, and the suggestion was made that the state agricultural society purchase him with the view of "introducing his superior blood into our state."[100] This animal was purchased, however, by a "public-spirited" citizen of Edgecombe County; in which place, it was announced, he would stand at stud.[101]

The Edgecombe County Agricultural Society sought to promote the local production of horses and mules by sponsoring a contest among citizens of that county who were engaged in raising those animals. The finest one and two year old colts were to be determined by judges who should base their selections primarily on form and other fine points. Persons who owned the two colts adjudged the "sorriest" were to furnish all the other exhibitors and judges with a barbecue dinner and all its trimmings.[102]

Notwithstanding the efforts made, the state failed to overcome completely its dependence upon outside sources for a considerable portion of its work stock. And of the stock produced in the state, by far the greater part was of the common breed—distinguished neither by beauty or other fine points. In the numbers of work animals owned, the state declined in relative rank in the Union in the 1850's, even though its relative position in the production of the major crops, cotton, tobacco, and rice, was being somewhat improved.[103]

Although the census reports and other contemporary sources of statistics failed to include any figures on poultry, there is no evidence that the population suffered from a scarcity in those items. Whether turkeys, chickens, guinea fowl, or ducks, they, like most other species of livestock, were thrown to a great extent on their own resources in foraging for a living and struggling for survival. Only in unusual cases was any shelter furnished them or nests provided for their accommodation. Just enough supplemental feed was given to keep them near the house and from reverting to a wild state. There was scarcely any market for surpluses of either eggs or meat, and poultry raising lacked that vital stimulus to improvement in practices.

Reformers urged poultry raisers to abandon their neglectful practices, to purchase and breed only improved fowl, and to make proper provision for their shelter and feeding. "The business of raising poultry

[98] *North Carolina Planter,* I (April, 1858), 117.
[99] *Ibid.,* 117.
[100] *Ibid.*
[101] *Ibid.,* I (May, 1858), 145, 160.
[102] Tarboro *Southerner,* November 25, 1854.
[103] *U. S. Census of 1860* (Agriculture), xxix ff.

and eggs is profitable," wrote the editor of the *North Carolina Farmer,* "but requires care, system and attention. You can't expect the hens to take care of themselves, lay eggs, hatch chickens, carry them to the market, and bring the money home to you."[104] "There seems to be no branch of domestic economy less understood than profitably raising poultry," wrote the editor of *The Soil of the South;* ". . . Most people think that chickens must pick up their own living."[105] Plans were suggested for those intending to build poultry houses and the revolutionary idea proposed that "They . . . will do as well shut up the year round as to run at large with proper care."[106] Cuts of various improved breeds of fowl, accompanied by descriptive articles, were carried in the agricultural press.[107]

The need for improvement in this branch of livestock raising was, perhaps, less appreciated than in any other, and very little was accomplished during the period beyond maintaining self-sufficiency.[108]

In practically every listing of prices current in the local markets, the items beeswax and honey appeared. Undoubtedly, great quantities of both were produced and honey was a very important sweet in the diet. The indices of production of these items, like those of nearly every crop, experienced a sharp rise in the latter part of the period. In the fifties the rate of increase amounted to a little more than 100 per cent.[109]

No significant developments in the culture of the bee were noted; however, reformers urged that they be properly housed in hives offering protection against the rain, ants, spiders, and other insects. "Every farmer should keep bees," advised one editor, "a few swarms to furnish honey for his own use, if no more. They toil with unremitting industry, asking but a full sweep of the wing, and no monopoly."[110]

In view of the gross carelessness and neglect with which all types of livestock were generally treated in the ante-bellum period, owners often not even seeing their possessions for months at a time, there can be but little wonder that diseases and pests exacted a heavy death toll, and often reduced the living to such a pitiful condition that the owners were "struck with horror at the sight."[111]

Having no veterinarian to apply to,[112] once disease had struck or the

[104] *North Carolina Farmer,* I (April, 1846), 260.

[105] Reprint in Tarboro *Press,* June 28, 1851.

[106] *Arator,* I (December, 1855), 274.

[107] For examples of the "Surry," "White Shanghai," "Buff Shanghai," and "Silver Pencilled Hamburg" breeds, see *Southern Cultivator,* XI (January, 1853), 24, (April, 1853), 125, (July, 1853), 211-212, (November, 1853), 349. This journal devoted a section to poultry, in which "premium fowls" were advertised and advice given on their care.

[108] E. Q. Hawk, *Economic History of the South,* 270.

[109] *U. S. Census of 1860* (Agriculture), 105-111.

[110] *Arator,* I (December, 1855), 274.

[111] *Farmer's Journal,* I (July, 1852), 102-103.

[112] There were only three veterinarians in the state in 1860. *U. S. Census of 1860* (Population), 363.

animal had become infested with parasitic pests, the owner, if in attendance at all, was thrown pretty much upon his own resources in his efforts to effect a cure or riddance. This situation gave rise to the frequent appearance of pleas for help which were published in about every issue of every agricultural paper. "Will you," wrote one farmer, "inform us of the best method of removing vermin from cattle, sheep, mules and horses—my sheep and cattle are very badly infested with them at this time."[113]

Ticks and lice were the "vermin" most frequently complained of, and numerous concoctions were evolved and publicized as being effective in securing their removal from the afflicted animal. Of the many ingredients used in the preparation of these remedies, brimstone, sulphur, tar, mercurial ointment, and turpentine were most common.[114]

Along with the recipe, good advice on the proper care and treatment of animals, similar to the following extract from the *American Farmer,* was often given:

> As an auxiliary to whatever remedy may be used, the curry comb and brush should be freely applied after a day or two, in order that the hyde and hair of the animal may be kept *clean.* No animal which is well fed, and daily curried and brushed, will either breed or retain lice; the latter operation, however, few who have much stock can regularly attend to.[115]

Diseases proved more difficult of diagnosis and treatment and generally caused great loss to stock raisers. Cholera in hogs and milk sickness in cattle were particularly destructive near 1860.[116]

Losses of livestock through theft or by defacing and obliteration of brands and markings continued heavy. The practice of allowing livestock to roam at large, unattended by the owner or his agent, and the infrequency with which the owner even saw his property, made such thievery a simple matter.[117]

In an attempt to safeguard raisers against such losses, a colonial law had prescribed severe penalties for these offenses, and new laws were enacted for the same purpose. In 1817, horse stealing was made a

[113] *American Farmer,* IV (February, 1823), 383. In reply to such requests, the editor either published a remedy or asked some reader to furnish one.

[114] For examples see *ibid.; North Carolina Farmer,* I (June, 1845), 15; *Farmer's Journal,* I (July, 1852), 102-103, 111; *ibid.,* I (August, 1852), 144; *ibid.,* I (October, 1852), 223; *Arator,* I (May, 1855), 62.

[115] Reprinted in *North Carolina Farmer,* I (June, 1845), 15.

[116] *North Carolina Planter,* II (August, 1859), 227; *ibid.,* III (September, 1860), 300; *ibid.,* I (August, 1858), 254.

[117] A correspondent from Harlowe's Creek, who desired to keep his communication "seacret," identified a neighboring tenant as a thief who would bear watching: ". . . when he rentid ther he onley had on[e] sheep and one cow and the sheap that he had was a ram an when he moved away he had 50 head of sheap which was a grate increas to raise from on[e] lone ram . . . when he left that plantation he had cattle enoug to find a poation of the united staits trupes in Beaf fore or five years. . . ." Unsigned, undated letter to J. W. Bryan, J. W. Bryan Papers, Southern Historical Collection, University of North Carolina Library.

"clergiable felony" for the first offense, and for the second, it was made a capital crime.[118] By an act of 1822, corporal punishment was prescribed as the penalty for intentionally altering or defacing brands or marks.[119] Thievery continued, nevertheless, since the laws against it were poorly enforced.

Undoubtedly, the most short-sighted and uneconomic aspect of ante-bellum North Carolina farming was the continued failure of agriculturists to produce sufficient forage for their livestock. Having become accustomed in colonial times, when settlement was sparse and the range was unlimited and abounding in rough forage, to allow livestock to roam at will and seek its subsistence where it could be found, stock raisers were extremely reluctant to change from that system after the conditions which made it profitable ceased to exist. The reasons for this reluctance are not hard to find. The range was free and the system of management commonly used in the state required outlays of only a minimum of capital and labor, of which, capital was scarce in ante-bellum North Carolina. Also, the profits upon investments continued high even after the scope of the industry became necessarily reduced.

The pioneer-like system of management of livestock, therefore, persisted and became the target of scathing criticism by the agricultural reformers and the more provident and intelligent stock raisers. Making, as it did, scarcely no provision for the shelter and supplemental feeding of animals, its evils were particularly apparent in the spring time. In speaking to the state agricultural society in 1853, A. W. Venable said:

> It is impossible to pass through the country in the spring without being pained to observe the cattle which have just achieved the enterprize of enduring the winter.
>
> Those which have survived, give unmistakable indications that their perils have been great, and the danger of starvation imminent.[120]

The Madison *News* reported in 1857 that many cattle were dying from starvation in that community because of the great scarcity of forage. "The most trifling straw," it was said, "either wheat, oat, or rye, cannot be had at any price. Another citizen, in order to save the life of a valuable milch cow, cut open a straw bed tick, and fed it away by piecemeals."[121]

Such conditions existed because the state had not yet become self-sufficient in the production of forage crops, a fact indicative of the livestock owner's general lack of intelligent self-interest in his holdings.

[118] *Laws of N. C.* (1817), 11.

[119] *Ibid.* (1822), 25.

[120] *Farmer's Journal,* II (November, 1853), 252-253.

[121] Reprinted in the Tarboro *Southerner,* May 16, 1857. The corn crop in 1856 was very poor and people as well as livestock suffered for lack of food during the winter and spring of 1857. See Tarboro *Southerner,* June 6, 1857; *ibid.,* June 20, 1857; *ibid.,* July 18, 1857.

"Our herds . . ." wrote an Edgecombe County farmer in 1853, "no longer thrive in the land, nor crowd the roads to market as in former times; for our wild-mast ranges are exhausted, and extensive new-grounds rarely offer their heavy winter foraging to our stock as formerly. The tide has turned back upon us. . . ."[122] To supply the deficiency, hay was imported into the state in considerable quantity for the feeding of livestock. Some even went to feed farmer's stock.[123]

The better farmers readily recognized the necessity for planting forage crops, and conducted numerous experiments with native and imported grasses and other crops. Between 1820 and 1860, the following were planted in the state for forage purposes: Guinea grass, feather grass, millet, Egyptian grass, lucerne, sainfoin, gama grass, herds grass or red top, timothy, crab grass, Hungarian grass, orchard grass, Chinese sugar cane, corn, oats, rye, pumpkins, turnips, sugar beets, carrots, peanuts, peas of various sorts, and several varieties of clover. Varying degrees of success were obtained. Lucerne or alfalfa, gama grass, herds grass, timothy, crab grass, orchard grass, corn, Chinese sugar cane, oats, rye, peas, and clovers showed greatest promise among the grasses.[124]

Henry King Burgwyn, of Northampton County, was one of the most successful large scale experimenters with forage crops, and his large stock of animals gave evidence of having been abundantly and well fed. Over a period of years, he planted timothy, herds grass, Kentucky blue grass, and great acreages of clover, and never failed to obtain good crops when the planting was done on good soil, properly prepared.[125] Burgwyn expended considerable labor and manures, especially lime, on his grass crops and concluded after ten years' experimentation that good permanent pastures could be made on any of the clay lands by such practice at far less cost than imported hay.[126] From the middle of May until the crop matured, he habitually grazed his stock of working horses in his clover fields without losing any of them; and, according to his observation, they improved in appearance and ability to work.[127] As for his other stock, "A Friend of the Dairy," who observed them, wrote

[122] *Farmer's Journal,* II (September, 1853), 181-182.

[123] *Proceedings of the Internal Improvement Convention, which was held at Salisbury . . . 1836,* 13-14; *Farmer's Journal,* I (September, 1852), 161-162; *ibid.,* II (November, 1853), 252-253; Tarboro *Southerner,* October 7, 1854.

[124] *American Farmer,* III (May, 1821), 53; *ibid.,* IV (December, 1822), 295-296; *ibid.,* VII (May, 1825), 58; *ibid.,* IX (February, 1828), 378; *Farmer's Register,* II (July, 1834), 123; *ibid.,* II (March, 1835), 616; *ibid.,* III (June, 1835), 126; Tarboro *Southerner,* October 7, 1854; *Arator,* I (May, 1855), 50; *North Carolina Planter,* I (June, 1858), 184, 190.

[125] Reprint from *American Farmer* in Tarboro *Press,* September 14, 1850. In 1850, Burgwyn had 220 acres of clover and grass, which had been sown the previous year, and planted an additional 315 acres in clover, herds grass, timothy, and rye grass. Reprint from Raleigh *Star* in Hillsboro *Recorder,* July 31, 1850.

[126] Reprint from *American Farmer* in Tarboro *Southerner,* June 25, 1853.

[127] Hillsboro *Recorder,* September 11, 1850.

that his beef, mutton, milk, and butter were the choicest to be found in the entire South.[128]

Culture of the grasses, whether for use as forage or for the enrichment of the soil, was emphasized as much, perhaps, as any other feature of the program for the improvement of agriculture. "Without the diligent and persevering cultivation of grasses," warned a speaker to the state agricultural society, "farming cannot prosper."[129] Paul C. Cameron would make the culture of grasses the foundation upon which to improve agriculture. He said:

> Could I be the means of inducing every farmer . . . to keep ten acres, of his best lands well set in clover and orchard grass, to be fed to stock in stalls, I should, in the production of manures, and in the increased means of living, have accomplished more for the solid wealth of the County, than an entire regiment of second class politicians.[130]

The *North Carolina Planter* employed Ebenezer Emmons, state geologist, to prepare a number of articles on grasses which were published in that paper. Professor Emmons had earlier given considerable attention to the native grasses in his geological reports and particularly emphasized them.[131] In this series, one of his first articles dealt with the common crab grass.[132]

Edmund Ruffin was greatly impressed by the luxuriant growth of peas, which he had observed in the state, and prepared a lengthy essay setting forth their value as a soil renovator and as a forage crop.[133] Ruffin thought the climate in North Carolina too hot to permit clover to attain its full growth, and for that reason, he believed peas better adapted for culture in the state. Where it could be grown, however, he highly recommended clover.[134]

Although immediate objection was raised to it because of the labor it required, the reformers recommended the "soiling" system of feeding livestock; that is, keeping the stock penned in stalls and feeding them on green forage which was cut and carried to them—allowing the stock out but for a short period of exercise daily. Ebenezer Emmons was the state's leading advocate of the soiling system. He would cultivate those grasses which came to maturity in succession for feeding in the growing seasons, and supplement the hay saved for winter feeding with generous quantities of sugar beets, turnips, carrots, and parsnips. Emmons recommended soiling to all those who wished to keep good stock; and, if properly done, he contended that it would be cheaper and better than

[128] Raleigh *Register*, April 1, 1857.
[129] *Farmer's Journal*, II (November, 1853), 252-253.
[130] *Arator*, I (June, 1855), 71.
[131] E. Emmons, *Report of the North-Carolina Geological Survey*, 135-191.
[132] *North Carolina Planter*, I (November, 1858), 337-338.
[133] E. Ruffin, *Essays and Notes on Agriculture*, 344-407.
[134] E. Ruffin, *Essays and Notes on Agriculture*, 125-149.

any other system.[135] Very few stock raisers, however, used the soiling system. It required too much labor for them and involved a too radical break with past practice to make its adoption easy.

All advocates of grass culture emphasized the importance of lime and plaster as manures to assist in the growth of these crops. As "fixed facts" of agriculture, the *American Farmer* advised that

> All lands on which clover, or the grasses, are grown, must either have lime in them naturally, or that mineral must be artificially supplied. It matters but little, whether it be supplied in the form of *stone-lime, oyster-shell* lime, or marl.
>
> All permanent improvement of lands must look to lime as its basis.[136]

Nevertheless, in spite of the many reasons that urged such a course, the production of forage crops did not increase in the proportion that they should have. Although the increase in the production of hay between 1840 and 1860 was about 80%, there still remained many evidences of the insufficiency of the forage supply.[137] In fact, some held that meadows were even more neglected than formerly.[138]

Some progress was made, however. Many forage crops had been introduced into the state and considerable improvement had been made in the feeding of livestock.[139] This was reflected by an increase in value of livestock and of animals slaughtered, during the decade of the fifties, by almost 100 per cent; whereas, the number of animals in the state had not increased by so large a percentage.[140] The people, too, had been given opportunities at fairs to see thoroughbred animals, and to draw their own conclusions as to the benefits to be obtained by selective breeding and proper care and feeding of livestock. By 1860, too, the state had overcome its deficiency in pork production and was shipping out more of that item than was being received. This was probably the case with beef, also. In the production of work stock—mules and horses—a greater percentage was being home grown, but the state continued to be partially dependent upon outside sources for its supply. More attention was being given to the proper breeding, sheltering, and feeding of livestock generally. As a whole, beginnings were made for the all-around improvement of livestock and methods employed in livestock management.

[135] E. Emmons, *Report of the North-Carolina Geological Survey,* 193.

[136] Reprinted in *Farmer's Journal,* I (August, 1852), 130.

[137] The increase was from 101,369 tons in 1840 to 181,365 tons in 1860. *U. S. Census of 1860* (Agriculture), 105-111.

[138] *Report of the Commissioner of Patents* (1850), 369; *North Carolina Planter,* I (March, 1858), 84.

[139] Benjamin M. Duggar, "Grass and Forage Crop Farming in the South," *The South in the Building of the Nation,* V, 227-228.

[140] *U. S. Census of 1860* (Agriculture), 105-111.

CHAPTER X

# AGRICULTURAL PROGRESS, 1783-1860: A SUMMARY

In summarizing the developments occurring in the agriculture of North Carolina during the period from 1783 to 1860, attention is again drawn to certain factors which exercised a significant influence over the trend of those developments. In the first place, the state as a whole suffered throughout the period from an inadequacy of transportation facilities, a deficiency which stood as an almost insurmountable obstacle to the development of its economic life. To a great extent, this condition was a result of unfavorable geographic circumstances. Because of the constant shifting of sand bars and other dangerous features along her coast line, mariners had found North Carolina's coastal waters inhospitable. Efforts to correct or to improve upon these conditions proved feeble and ineffective. At no time during the period did the state have an adequate commercial outlet to salt water. Neither were her rivers suited to navigation, except to a limited extent in the eastern part of the state. For shipping, resort was had principally to the ports of Virginia and South Carolina.

These natural disadvantages were but little offset by the state's internal transportation system. Following the pattern established in the colonial period, the state persisted in shifting the responsibility for the construction and upkeep of public roads upon the communities through which they passed. As a result, there was not a single, suitable all-weather public road in the state prior to 1860. Trade and commerce suffered from the lack of these vital facilities, and were, therefore, necessarily restricted. The numerous toll gates, bridges, and ferries added to the already heavy costs of travel and transportation. No cities developed to offer facilities for trade or as an adequate local market for surplus agricultural produce. For the greater part of the period, most of the state was considerably isolated from intercourse with the outside, and was inaccessible except at tremendous cost in money, labor, and comfort. It was not until after 1835 that these handicaps were partially overcome by plank roads and railroads.

Transportation costs continued heavy, almost too heavy. This situation kept the state in a particularly disadvantageous position in competition with other states. A Granville County producer, for example, had to pay as much to transport a hogshead of his tobacco sixty miles over the Petersburg Railroad to market in that city as a

Missourian paid to transport one of his hogsheads five hundred miles to the New Orleans market.[1] Granville County had the good fortune of being on the route of the Petersburg Railroad. For those North Carolina communities which were not so fortunate, the difference in the per mile costs of transportation was even greater. Such communities found it extremely difficult to compete with those having adequate transportation.

Principally because of the remoteness from markets and the heavy transportation costs, ante-bellum North Carolina agriculture tended more towards self-sufficiency and less towards commercial planting than perhaps was the case with any of the other slave-holding states. In the production of staple crops, the margin of profits for the average planter was so low that, of necessity, he had to produce his own requirements of subsistence crops. That some maintenance and subsistence items, such as corn, flour, forage, livestock, and livestock products, were imported into the state is not a denial of the above, but simply an indication of bad planning, short crops, or a poor distribution of local supplies resulting from inadequate transportation facilities. Failing in the production of their own requirements of wheat, for example, the people of Newbern could buy flour which had been grown and milled in New York state more cheaply than they could the surpluses that lay in Rowan County granaries. Economically, if not in keeping with the sectional political philosophy, it was the thing to do. In the state as a whole, however, the production of the requirement of subsistence crops held a priority of interest with North Carolina agriculturists. The variety of soils and climate in the state also encouraged diversification. That substantial success was had in this direction is indicated by the fact that, at the end of the 1850's, a decade in which more attention was given to the production of commercial crops than in any other, the per capita value of food crops produced in North Carolina exceeded the average for the entire nation as well as that of each of her immediate neighbors.[2]

The extent to which the state could participate in the production of the staples, particularly cotton and rice, was dictated by her geographic setting and circumstances. Laying as it did athwart the northern limits of the region in which those crops thrive, much of the area of the state was denied to their culture. The profits in the cultivation of tobacco remained so low or uncertain through so long a time that culture of the crop did not reach its broadest expanse in the state by 1860. Scarcely any ante-bellum North Carolina planters concentrated upon the production of a single staple. They chose instead to produce a variety of crops with the result that, in 1845, Governor William A.

[1] *North Carolina Farmer,* III (September, 1847), 78.
[2] *U. S. Census of 1860* (Agriculture), 104-111.

Graham said it was difficult to say which of the state's export crops was of greatest value: cotton, tobacco, rice, wheat and flour, or corn.[3]

Very little change occurred during the period in either the composition or economic interests of the population. Agriculture continued to be the principal source of a livelihood for an overwhelming majority of the people. More than in any other state, this population was native born. The state gained proportionately fewer people by immigation during the period and lost more by emigration than any other state in the Union.[4] The census of 1860 recorded over one-quarter of a million natives of the state who were living elsewhere. In this respect, North Carolina's contribution to the growth and expansion of the Union was second to none. The exodus of farm people, which continued to 1860, not only reflected upon the character and prospects of North Carolina agriculture but tended to retard its improvement. Among those who remained in the state, the Negroes increased in numbers at a faster rate than the whites. In 1790, white people made up 73.19 per cent of the population, and, in 1860, only 63.46 per cent. In relative rank in total population the state declined from fourth place in 1790 to twelfth in 1860.[5]

Spread out as it was, the population averaged only nineteen per square mile in 1860, the average ante-bellum North Carolina farmer lived in isolation from his neighbors. Having poor roads and means of communication, the social life of the people was necessarily restricted. Such circumstances fostered the growth of prejudice, superstition, and narrowness, and, a continuation of practices already outmoded or disproved in more enlightened neighborhoods. Association of the people for common betterment in schools, churches, clubs, and other community activities was difficult. Civic or community pride was practically nonexistent. Farming became and tended to remain a customs-ridden business. Attitudes of mind were developed under such conditions which stood as unreasoning bulwarks of opposition to change. Initiative was throttled or scoffed at. North Carolinians acquired a "proverbial reputation" for their "ignorance and torpidity."[6] To a Nash County emigre of this period, they were "farther behind the age than any people in the United States, and are more difficult to be aroused to a sense of their vital interests. . . ."[7] Illiteracy rates were high. In 1840, there were 56,609 adult whites over twenty years of age living in the state who could not read or write. This number was exceeded only in Virginia and Tennessee. In 1850, the number had increased to 73,566

[3] *Legislative Documents* (1844-1845), 7.
[4] *DeBow's Review*, XVII (October, 1854), 431.
[5] *U. S. Census of 1860* (Statistics), xx.
[6] F. L. Olmsted, *Journey in the Seaboard Slave States*, 366.
[7] *Farmer's Journal*, III (May, 1854), 53.

and was again only exceeded by the same two states and New York.[8] Due principally to the improvements in the public school system, the number of illiterates in 1860 was reduced to 68,308.[9]

Under such circumstances, there can be but little wonder that outmoded patterns of conduct and practice persisted in the population and could only be changed with great difficulty. This reluctance to change and improvement was particularly noticeable in, and harmful to, the state's agriculture. In some respects, farming was unbelievably crude and unintelligent. Lands were scratched over and planted year after year in the same crops without any provision being made for restoring the fertility extracted from the soil, or to prevent its total wastage by erosion. The land itself was not regarded highly by the tillers. Land was the plentiful factor. The average ante-bellum North Carolina farmer expended a minimum of labor and capital on a maximum number of acres. This resulted in low yields per acre, increasing production costs, soil exhaustion and wastage, and ultimate abandonment by the farmer. This chain of developments, along with the inadequacy of transportation and marketing facilities, served as the principal stimuli to keep the tide of emigration moving. Inclined, as the average farmer was, to ridicule "book farming" as being a humbug, he opposed changes and innovations of all sorts, and clung tenaciously to methods which actually marred the character of agriculture and stamped the practitioners as ignorant and unprogressive. Such a farmer, and ante-bellum North Carolina had many of them, had little or no interest in agricultural education, farm journals, agricultural societies, and the like. To him, they were humbugs also. He, and those like him, did not subscribe to the farm journals and all seven of those published in the state failed for the lack of adequate subscriber support.

In spite of the truth of such generalizations, there were in the state some agriculturists who ranked with the most progressive in the land in the degree of intelligence they applied to farming. They were not many, but their number constantly increased. Due principally to their experimentation, enterprise, and example, the downward course of agriculture in the state was arrested; and, after about 1835, a great improvement in farming practices became noticeable. This accomplishment must be rated the most significant development in agriculture during the period. There were proportionately more "good" farmers in the state in 1860 than ever before. This fact is substantiated by the indices of increasing production contained in the censuses of 1840, 1850, and 1860; by the great increase in the value of farm property; and, by the greatly decreased flow of emigration.

As a final factor which influenced developments in agriculture, the role of the state must be mentioned. In accordance with the political

[8] *Seventh Census of the United States: 1850,* lxi.
[9] *U. S. Census of 1860* (Statistics), 508.

philosophy prevailing at the time of its founding, the state government from 1783 to 1835 manifested only a small interest in the problems that confronted the people. In spite of the obvious need of good public roads and other internal improvements which would enable farmers to bring their surpluses to market at a profit, the state government took little action prior to 1835 to correct that condition. Neither were public schools established, and, as a result, a great proportion of the population grew up in ignorance. With such a citizenry, superstitious beliefs found easy fixation and stood as obstacles to the institution of a scientific agriculture. Under such circumstances, there can be little wonder that the state's agriculture became run-down and unprosperous. After 1835, when the constitution was reformed, there also came a change in the spirit of the state government. Between that date and 1860, several significant measures were taken by the state which contributed towards revival and reform in its economic life. Of greatest consequence to agriculture, were the establishment of public schools, and the construction of internal improvements, particularly railroads, of which, the state had a total of 890 miles in 1860.

Fortunately, there were in North Carolina throughout the period some agriculturists who were alert to the developments occurring in farming elsewhere, and who were experimenting with the view of instituting improvements in their own practices. To these men, credit must go for beginning and popularizing changes in the agricultural methods. Culture of crops in hills was almost abandoned, and, by 1860, they were nearly all being planted in rows or broadcast. The benefits of horizontal or contour plowing became widely known, and were put into use in efforts to conserve fertility and prevent erosion of the top soil. Better drainage of crop lands was effected and farmers began to acquire more knowledge as to the moisture requirements of various crops. Deep plowing was used in the preparation of the soil, and a more careful and thorough culture employed with the growing crops. Farmers began to acquire information as to the properties of the soil, and more of them undertook to stimulate its productivity by applications of manures and fertilizers. Not all farmers used these improved practices, of course, but those who did greatly increased per acre yields and improved the quality of their products.

Development and invention of improved or new farming implements, in which North Carolinians had a hand, greatly increased the productivity and efficiency of agricultural workers. The hoe and the mattock, although by no means displaced, lost importance to the cultivators and iron and steel plows which were brought into widespread use. Likewise the sickle gave way to the mowing scythe, which, in turn, was antiquated by the reaper. All three, however, continued in use in North Carolina. One such invention called for another, and the thresher was invented. Horsepower machines, fanning mills, drills, planters, manure carts, corn shellers, corn pounders, hay rakes, cotton gins, cotton presses, and wagons were either invented or improved. Slave workers demonstrated equal ability with free men in the use of these new implements.

In the realm of crops, the development of "bright" tobacco had greatest portent for the future of the state. In other respects, the culture of tobacco, like that of rice and cotton, experienced a varying emphasis as a money crop. Cotton culture, which had spread rapidly over the state after 1800 and then declined rapidly to the point where it had almost been abandoned as a money crop by 1835, revived just as rapidly after 1840 under the stimuli of improved transport and marketing facilities, more intelligent methods of culture, and higher prices. In 1860, cotton was the leading money crop of the state. Rice, on the other hand, although increasing in production during the fifties, became more and more confined in its area of growth to the far southeastern portion of the state. Corn, wheat, and the other small grains benefited from the improvements in the implements and methods of culture but continued to be produced principally as subsistence crops. The same may be said of Irish potatoes, sweet potatoes, and beans. Peanuts, introduced late, but found to be a highly productive and satisfactory crop, were becoming a popular crop by 1860. The clovers, peas, and grasses, whether planted for forage or soil building purposes, were being much more widely cultivated.

The commercial nursery made its initial appearance in the state and was an important development in its agriculture. Numerous varieties of fruit trees, vines, shrubs, berries, and vegetables were selected, propagated, tested, and offered for sale. Fruit and vegetable growing was stimulated. A few native varieties, such as the Catawba and Scuppernong grapes, became widely popular in the nation. The state won the distinction in 1840 of being first in the Union in the production of wine. Apples grown in the state won highest honors in national exhibitions. "I have the means of knowing," said Daniel M. Barringer in 1859, "that during the last year, the cash return for fruit grown in this state, was nearly a half million dollars."[10] Of greater importance than the fact that a new source of income was developing for the state, was the fact that an abundance of fruits was being produced to meet the people's needs. North Carolina fruit growers also had a significant part in modifying European practices in fruit culture to suit the different growing circumstances which prevailed in America.

In relative standing with other states in the raising of livestock, particularly beef cattle and swine, the state experienced considerable decline from the lofty position it held in 1783. This may be attributed

[10] D. M. Barringer, *Address Delivered Before the Mecklenburg Agricultural Society*, 17-18.

to two principal causes: in North Carolina the common grazing lands were being constantly reduced in extent by the extension of settlement; and, in other states, it was found that by selective breeding and proper feeding better animals could be produced than those purchased previously from North Carolina. The state, therefore, declined from the position of seller to that of buyer of cattle, swine, sheep, horses, and mules, and, in spite of efforts directed toward the production of her own requirements of these animals, some were still being purchased in 1860 from outside the state. The volume of such purchases, however, had been greatly reduced by that date.

In other respects the period was notable for its developments in the realm of livestock. Mules were introduced into the state and proved popular from the beginning. By 1860, they were rapidly replacing the horse for all farm purposes except hauling, draying, riding, and for carriage purposes. North Carolinians also had a part in the craze for Merino sheep which swept over the country about 1810. Sheep raising, however, never attained commercial importance in the state. Many North Carolina stock raisers sought to improve the quality of their animals, whether horses, mules, sheep, swine, milk cattle, beef cattle, or poultry, by cross-breeding them with highly blooded stock which they had imported for the purpose. The number of such animals imported was not large, but in this manner the blood of the best animals in the land was mixed with that of the native breeds. By 1860, undoubtedly a considerable improvement in the quality of livestock had been made.

Between 1783 and 1860, too, the American agricultural press had its beginning. Prior to the publication of any farm journal in either the state or nation, several North Carolina newspapers began devoting space to agricultural subjects. Most of the articles were reprints from out-of-state papers and quite often showed unwise selection and the need of editing, but in this manner a great volume of information on farming was disseminated among the people. Beginning with the publication of the *American Farmer,* in 1819, more and more agricultural journals were published, and, undoubtedly, they exercised a great influence on developments occurring thereafter. With the more intelligent agriculturists, "book farming" came into better repute.

Enterprising publishers were also encouraged to attempt the publication of farm journals in North Carolina. Between 1838 and 1860, seven different papers were published in the state; but, although some of them, including the *Farmer's Advocate,* and the *Farmer's Journal,* possessed real merit, none of them received adequate financial backing or a sufficiently large number of subscribers to continue publication longer than four years. These ventures must not be written off as complete failures, however. Their pages are filled with testimonials of farmers who gave these papers credit for bringing new hope to the state's agriculture.

Numerous volumes on agricultural subjects also made their appearance and were distributed, at nominal prices, from the book stores in the state. The only one of significance by a North Carolinian was George W. Jeffrey's, *Series of Essays on Agriculture and Rural Affairs; in Forty Seven Numbers,* which was brought out in 1819, and which echoed the teachings of John Taylor's better known *Arator.* Circulation of this sort of literature, however, was seriously limited by the widespread prejudices against anything that might be labeled "book farming" which persisted among the mass of farmers. With most such farmers, the almanac fulfilled their secular reading requirements. Several of these were published in the state and substantially supported by a large volume of subscribers. These, too, carried a few articles on agricultural subjects.

The organization of agricultural societies, which came as a new development, gave great promise for the future promotion of the farm interests. Although these societies failed to attract the support of the great mass of farmers, and quite often failed to persevere in their own purposes, they did stimulate and encourage better farming by focusing the light of inquiry upon current farm practices, by promoting the acceptance of new ideas, by encouraging experimentation, and by promoting agricultural education, the agricultural press, agricultural fairs, and stock shows. Members began to take more pride in their profession, and to rear their children to be farmers. Rural life was pictured as the purest of all, and the farmer as the worker who stood highest in favor with God.[11] Encouraged by the obvious improvements made in farming and by the financial support given by the state, these societies were more active in promoting good farming than any other agencies in the state.

More information concerning the chemical properties of the soil and their functions in the production of crops was gathered than in any previous time. As this data became disseminated, the more intelligent agriculturists began to realize that theirs was a business that required not only common sense, but scientific knowledge as well. Out of this realization sprang the demand for agricultural education. If farming was a business dependent for its success upon the presence of certain elements in the soil, a complete survey of the state's agricultural resources by a competent scientist was essential. To accomplish these purposes, a professorship in agriculural chemistry was established in the University of North Carolina, and a second geological survey of the state was ordered, with particular reference to its agricultural potentialities. These developments, however, came so late that their influence on agriculture was only beginning to be felt in 1860.

In general education, the advance was much more rapid. By 1860,

[11] D. M. Barringer, *Address Delivered Before the Mecklenburg Agricultural Society,* 26-31.

the public school system, which was made possible by the spirit of reform which swept over the state after 1835, was operating with great success. The rate of adult illiteracy was being gradually lowered and the pall of ignorance lifted from the land. With the promise this development offered for the eradication of prejudice, superstition, and narrowness, and for the encouragement of ambition, progress, and civic pride, the promoters of an improved agriculture could look forward to a better day. It was early demonstrated that public schools and agriculture would be mutually beneficial. Some of the swamp lands which had been allocated to the literary board as a source of funds for the establishment of public schools had, by 1860, been drained, sold, and brought into rich agricultural production.

Beginning with the census of 1840, each succeeding decennial census devoted more and more space to information concerning the nation's agriculture. Under a large number of headings, data for these volumes were secured from each farmer, totaled by counties, states, and sections, and published to all the nation. In these published tables, North Carolinians found several items pertaining to the state's agriculture in which they could take pride. This was a new experience. The state was not at the bottom of the list in everything as some citizens had undoubtedly been led to believe. Actually, in several particulars North Carolina had out-distanced her more prideful neighbors. This information was disseminated throughout the state and served to elevate the morale of the farm population.[12]

Furthermore the reports on agriculture, which were compiled annually after 1841 by the United States Commissioner of Patents, became increasingly voluminous and undoubtedly proved helpful to those who used them. A great proportion of the information contained in these volumes was procured by circularizing representative farmers in all the states. Likewise, farmers who had gained distinction in some aspect of their operations were given opportunity to publish accounts of their successes in these volumes. North Carolina contributors are to be found in both categories. These reports were widely circulated throughout the land, and mark the initial appearance of the federal government in the field of agricultural promotion.

The weight of evil circumstances brought the relative standing of North Carolina agriculture to its lowest point in the 1830's. From the latter part of that decade unil 1860 conditions improved. Advanced methods began to find acceptance among a larger number of farmers. Extensive and soil exhausting practices began to give way to intensive and soil conserving modes of culture. Better implements were sought out and employed. Some farmers began to use high quality work animals and other livestock. The state assisted agriculture by taking

[12] *Farmer's Journal,* III (November, 1854), 240-242.

great strides towards solving the problems of transportation and marketing, and, with this development, the North Carolina farmer's ability to compete in the production of the staples was enhanced. Conformity with other states in standards of weight, measure, and quality of agricultural produce was sought. Prices generally were higher after 1840 and agriculture again became a profitable business for North Carolinians. A more progressive and optimistic spirit became observable in the population. Emigration declined in the last two decades to a mere trickle. Land values rose tremendously, and, in the 1850's, North Carolina agriculture enjoyed greater prosperity than ever before.

This reformation and upsurge in the state's agricultural fortunes is reflected most clearly, perhaps, by the statistics of increased production between 1840 and 1860. Of the staple crops, cotton production increased by about 350 per cent in the two decades; tobacco by 100 per cent; and, rice by about 268 per cent. Of crops grown primarily for subsistence purposes, corn increased by approximately 30 per cent; wheat by 248 per cent; and, sweet and Irish potatoes combined by 265 per cent. Several of the lesser crops also showed significant advances. The following table illustrates the advances made in production:[18]

PRODUCTION OF CROPS IN NORTH CAROLINA, 1840, 1850, AND 1860

| Crop | 1840 | 1850 | 1860 |
|---|---|---|---|
| Cotton | 51,926,190 pounds | 73,845 bales | 145,514 bales |
| Tobacco, lb. | 16,772,359 | 11,984,786 | 32,853,250 |
| Rice, lb. | 2,820,388 | 5,465,868 | 7,593,976 |
| Wheat, bu. | 1,960,855 | 2,130,102 | 4,743,706 |
| Corn, bu. | 23,893,763 | 27,941,051 | 30,078,564 |
| Potatoes, bu. | 2,609,239 | ......... | ......... |
| Potatoes, Irish, bu. | ......... | 620,318 | 830,565 |
| Potatoes, Sweet, bu. | .... .... | 5,095,709 | 6,140,039 |
| Peas and Beans, bu. | ......... | 1,584,252 | 1,932,204 |

These were the accomplishments in a twenty year period! Loyal sons of the state began to boast that "Old Rip" had at last awakened from his slumbers.

The statistics of the 1840 census are less complete than those of 1850 and 1860, and, therefore, it is impossible to measure the progress, in all particulars, of the agricultural advances being made. Nevertheless, between 1850 and 1860 the cash value of farms in the state increased by

[18] *Compendium . . . From the Returns of the Sixth Census. . . .*, 178-180; *The Seventh Census of the United States: 1850. . . .*, 320-322; *U. S. Census of 1860* (Agriculture), 105-111. A prolonged drought in the Piedmont counties in 1859 (the year agricultural statistics were taken for the census of 1860) severely reduced crops in that section. Corn production was particularly hard hit, in some counties not as much as one-half a crop being made. *North Carolina Planter*, II (August, 1859), 243.

slightly more than 100 per cent, which rate was just above that for the nation as a whole. In the same decade, the value of farm implements and machinery increased by sixty-seven per cent; the value of livestock by eighty-two per cent; and, the value of animals slaughtered by eighty-two per cent. Between these two dates, the agriculture of the state registered advances in production in every branch of the industry except the following: sheep, wool, oats, cheese, clover seed, hops, flax, and flaxseed. As might be expected of a state so predominantly agricultural, the greater emphasis placed upon that interest, caused certain other interests to show corresponding decreases. The value of homemade manufactures, for example, declined slightly between 1850 and 1860.[14] The "true value" of real and personal property of all sorts, however, increased during the decade by fifty-eight per cent.[15]

As interest in staple production grew, certain counties which were more or less specializing in their culture began to exhibit more of the tendencies of the plantation regime. In Edgecombe County, for example, between 1840 and 1860, the white population declined by thirteen per cent, whereas, the slave population increased by thirty-five per cent. In the same period, the production of cotton, the principal staple, increased by over 900 per cent. Production of subsistence crops continued fairly constant, but the county lacked self-sufficiency in work stock and forage.[16] Land values rose tremendously. Likewise, the price of slaves went up. These were but the signs of a more prosperous agriculture.

Counting slaves, North Carolina had a total population in 1860 of 992,622, and ranked twelfth in the Union in that respect. Based on per capita production, the state's relative standing in the Union had been considerably improved by that date. This fact is indicated in the following table:[17]

Although the state made remarkable progress in increasing agricultural production between 1840 and 1860, yet, the rate of advance was in many instances not as fast as that being made in the nation at large. North Carolina, therefore, declined in relative standing in the production of those items. It was only in the case of sweet potatoes, rice, tobacco, and swine that the state's relative position as a producer was improved between those dates.[18] On a strictly per capita basis, the position of North Carolina was more favorable. In 1860, the state's rank in the production of the following items was equivalent to its

[14] *U. S. Census of 1860* (Agriculture), 184-192. In the nation at large, the value of farm implements and machinery increased by only sixty-three per cent. *Ibid.*, xv.

[15] *Ibid.* (Statistics), 295.

[16] *Compendium . . . From the Returns of the Sixth Census. . . .*, 40-43, 179-181; *Seventh Census of the United States: 1850*, 307, 318-321; *U. S. Census of 1860* (Agriculture), 104-111; *ibid.* (Population), 348-355.

[17] *U. S. Census of 1860* (Agriculture), 104-111.

[18] *Compendium . . . From the Returns of the Sixth Census*, 358-364; *U. S. Census of 1860* (Agriculture), xxix ff.

PER CAPITA PRODUCTION OR OWNERSHIP OF VARIOUS AGRICULTURAL ITEMS—1860

| *Item* | *United States per capita* | *The South per capita* | *North Carolina per capita* |
|---|---|---|---|
| Horses, asses, and mules | 0.24 | 0.29 | .21 |
| Sheep and swine | 1.78 | 2.27 | 2.74 |
| Cattle-dairy and oxen | 0.81 | 1.05 | .81 |
| Wheat, rye, barley and buckwheat, bu. | 7.24 | 4.57 | 5.26 |
| Corn, bu. | 26.68 | 35.63 | 30.32 |
| Irish and sweet potatoes, bu. | 4.87 | 4.23 | 7.02 |
| Peas and beans, bu. | 0.48 | 0.98 | 1.94 |

ranking as the twelfth most populous state: corn, peas and beans, sweet potatoes, flax, cotton, rice, tobacco, asses and mules, and swine.[19]

Notwithstanding the fact that the rate of agricultural progress in the state was not fast enough to keep up with the nation in some respects it was rapid enough to excite the enthusiasm of some of the promoters, planters, and public officials. "A new impulse has been given this branch of industry," observed Governor David S. Reid to the state legislature in 1854, "that is perceptible in every part of the state."[20] "Our great interest, agriculture," reported Governor Thomas Bragg to the same body in 1856, "to say nothing of the others, is attracting the notice it deserves, and our people are on the enquiry for the best modes of improving their lands and increasing their crops."[21] Doctor John F. Tompkins, editor of the *Farmer's Journal,* wrote in 1854 that North Carolina "has at last aroused from her slumbers, shaken off her dull feelings and entered with her sister States, fully and fairly upon a thorough improvement in agriculture."[22] John S. Dancy wrote from Edgecombe County of the "wonderful revolution in the agriculture of our state," and ascribed "to Mr. Ruffin . . . the compliment of having brought it about."[23] Daniel M. Barringer told the farmers of Mecklenburg County in 1859 that "The fatal causes of a pernicious deterioration are no longer in operation. Emigration has nearly or quite ceased to drain us of our people and our wealth. . . . our lands throughout the state have more than doubled [in value]—and in this immediate section, have tripled and quadrupled in value in the last few years."[24] Ebenezer Emmons cited the improvements in the state's transportation and marketing facilities as being chiefly responsible for the upsurge of

[19] *U. S. Census of 1860* (Agriculture), xxix ff.
[20] "Executive Document No. 1," *Legislative Documents* (1854-1855), 9.
[21] "Executive Document No. 1," *Legislative Documents* (1856-1857), 3.
[22] *Farmer's Journal,* III (October, 1854), 208.
[23] *Arator,* I (August, 1855), 138.
[24] D. M. Barringer, *Address Delivered Before the Mecklenburg Agricultural Society,* 7.

agriculture.[25] From all over the state came testimonials recounting the progress being made. On nearly every hand, there were indications that the farmer was prospering and that his morale was up.

Although many farmers clung tenaciously to the outmoded practices of their fathers, the agriculture of the state showed many evidences of new developments and that it was in process of being reformed and stimulated. A customs-ridden business to begin with, it is not surprising that agriculture made slow progress and that the number participating in the reform movement was small. Farmers generally are individualists and strongly inclined towards conservatism. Ante-bellum North Carolina farmers possessed such characteristics in full measure. In spite of all obstacles, however, by 1860 the foundations of a better agriculture had been firmly established in every community in the state.

[25] E. Emmons, *Agriculture of North Carolina, Part II,* vi.

# BIBLIOGRAPHY

## I PRIMARY SOURCES


### A Manuscripts

Burwell Benson Papers, 1833-1922. Southern Historical Collection, University of North Carolina Library. The parts of this collection used deal with home life and economic conditions on a small Alamance County farm, 1853-1857.

Bryan Papers, 1704-1929. Southern Historical Collection, University of North Carolina Library. The correspondence of J. W. Bryan during the period from 1830 to 1860 was used principally. This was concerned with business and agricultural affairs of Carteret, Craven, Lenoir, Jones, Onslow, and Wayne counties.

Henry King Burgwyn Papers, 1764-1877. Southern Historical Collection, University of North Carolina Library. These papers deal principally with the correspondence of a planter with his overseer on such matters as slave management, agricultural methods, farm machinery, and general plantation management.

Stephen Andrews Norfleet Papers, 6 vols. Southern Historical Collection, University of North Carolina Library. This source consists chiefly of the diary and account books maintained by Stephen A. Norfleet, of Bertie County, during the 1850's. Entries covering practically every phase of plantation operations are included in the diary.

Peter Evans Smith Papers, 1739-1925. Southern Historical Collection, University of North Carolina Library. Papers used from this collection had to do with cotton prices, and the operations of a cotton farmer in Edgecombe County in the 1850's.

Lewis Thompson Papers, 1717-1894. Southern Historical Collection, University of North Carolina Library. This is a particularly valuable source of information on the operation of a plantation in Bertie County from 1835 to 1868. Problems connected with the cultivation of cotton, grains, and other subsistence crops are dealt with, and, also, the management and hire of slaves, the purchase of tools, and correspondence with commission merchants.

### B Official Documents

*Agriculture of the United States in 1860; Compiled from the Original Returns of the Eighth Census.* Washington: Government Printing Office, 1864.

*Annual Report of the Board for Internal Improvements of North Carolina.* . . . (1823). Raleigh: J. Gales & Son, State Printers, 1823.

*Annual Report of the Board of Public Improvements of North Carolina. . . .* (1821). Raleigh: Printed by J. Gales, 1821.

Clark, Walter, ed., *State Records of North Carolina,* 20 vols. Winston and Goldsboro: Nash Brothers, 1895-1906.

*Compendium of the Enumeration of the Inhabitants and Statistics of the United States as Obtained . . . From the Returns of the Sixth Census.* Washington: Printed by Thomas Allen, 1841.

Curtis, Moses Ashley, *Geological and Natural History Survey of North-Carolina, Part III, Botany: Containing a Catalogue of the Plants of the State, with Descriptions and History of the Trees, Shrubs, and Woody Vines.* Raleigh: W. W. Holden, Printer to the State, 1860.

DeBow, James Dunwoody Brownson, *Statistical View of the United States . . . Being a Compendium of the Seventh Census. . . .* Washington: A. O. P. Nicholson, Public Printer, 1854.

Emmons, Ebenezer, *Agriculture of North-Carolina, Part II; Containing a Statement of the Principles of the Science upon which the Practices of Agriculture, as an Art, are Founded.* Raleigh: W. W. Holden, Printer to the State, 1860.

Emmons, Ebenezer, *Geological Report of the Midland Counties of North Carolina.* Raleigh: Henry D. Turner, 1856.

Emmons, Ebenezer, *Report of the North-Carolina Geological Survey. Agriculture of the Eastern Counties; Together with Descriptions of the Fossils of the Marl Beds.* Raleigh: Henry D. Turner, 1858.

Emmons, Ebenezer, *The Swamp Lands of North Carolina.* Raleigh: W. W. Holden, Printer to the State, 1860.

*Executive Document,* No. 13., "Report of Professor Emmons, on his Geological Survey of North Carolina." Raleigh: Seaton Gales, Printer to the Legislature, 1852.

Hilgard, Eugene W., "Report on Cotton Production in the United States. . . . ," Part II, *Tenth Census of the United States.* Washington: Government Printing Office, 1884.

*Journals of the Senate and House of Commons of the General Assembly of North Carolina,* 1848-1849.

Killebrew, Joseph Buckner, *Report on the Culture and Curing of Tobacco in the United States.* Washington: Government Printing Office, 1884.

*Laws of North Carolina,* 1810-1860.

*Laws of the State of North Carolina, Including the Titles of Such Statutes of Great Britain as are in Force in Said State; together with . . . Marginal Notes and References. . . .*, 2 vols. Raleigh: Printed and sold by J. Gales, 1821.

*Legislative Documents,* 1829-1861.

*Manufactures of the United States in 1860; Compiled from the Original Returns of the Eighth Census.* Washington: Government Printing Office, 1865.

Mitchell, Elisha, *Report on the Geology of North Carolina, November 29, 1827.* North Carolina, Board of Agriculture, *Papers,* III, 1-27.

North Carolina, Board of Agriculture. *Papers on Agricultural Subjects,* 4 vols. Raleigh: J. Gales & Son, 1824-1829.

Olmsted, Denison, *Report on the Geology of Part of the Western Counties of North Carolina.* North Carolina, Board of Agriculture. *Papers,* I-II. 1825-1827.

*Population of the United States in 1860; Compiled from the Returns of the Eighth Census.* Washington: Government Printing Office, 1864.

*Report on Internal Improvements in North Carolina, 1819-1820.* Raleigh: Printed by Thomas Henderson, 1819.

*Report of the Board for Internal Improvements, 1833.* Raleigh: Printed by Lawrence & Lemay, 1833.

*Report of the Commissioner of Patents,* 1843-1860.

*Report on the Productions of Agriculture as Returned at the Tenth Census.* Washington: Government Printing Office, 1883.

*Reports to the General Assembly on Public Improvements in North Carolina, 1822.* Raleigh: Printed by J. Gales & Son, 1822.

*Return of the Whole Number of Persons within the Several Districts of the United States, . . . .* (Census of 1790). Washington City: Printed by William Duane, 1802.

*Return of the Whole Number of Persons within the Several Districts of the United States, . . . .* (Census of 1800). Washington City: Printed by William Duane & Son, 1802.

Saunders, William Laurence, ed., *Colonial Records of North Carolina,* 10 vols. Raleigh: P. M. Hale, [etc.] State Printer, 1886-1898.

*Statistics of the United States (Including Mortality, Property, & c.) in 1860; Compiled from the Original Returns and Being the Final Exhibit of the Eighth Census.* Washington: Government Printing Office, 1866.

*Tenth Census of the United States, 1880, VI, Report on Cotton Production in the United States. . . . ,* Part II. Washington: Government Printing Office, 1884.

*The Cotton Plant: Its History, Botany, Chemistry, Culture, Enemies, and Uses.* 54th Congress, 2nd Session. House Document No. 267. Washington: Government Printing Office, 1896.

*The Seventh Census of the United States: 1850. . . .* Washington: Robert Armstrong, Public Printer, 1853.

C Collected Documents and Contemporary Writings

Allen, John Fisk, *A Practical Treatise on the Culture and Treatment of the Grape Vine: Embracing Its History, with Directions for Its Treatment, in the United States of America, in the Open Air, and Under Glass Structures, With and Without Artificial Heat,* 3rd. ed. New York: A. O. Moore, Agricultural Book Publisher, 1859.

Allen, Richard Lamb, *The American Farm Book: or Compend of American Agriculture: A Practical Treatise on Soils, Manures, Draining, Irrigation, Grasses, Grain, Roots, Fruits, Cotton, Tobacco, Sugar Cane, Rice, and Every Staple Product of the United States with the Best Methods of Planting, Cultivating, and Preparation for Market.* New York: C. M. Saxton, 1860.

Barringer, Daniel Moreau, *Address Delivered Before the Mecklenburg Agricultural Society.* Charlotte: Printed in the Office of the *Western Democrat,* 1860.

Bassett, John Spencer, ed., *The Southern Plantation Overseer, As Revealed in His Letters.* Northampton, Massachusetts: Printed for Smith College, 1925.

Bassett, John Spencer, ed., *The Writings of Colonel William Byrd.* New York: Doubleday, Page and Company, 1901.

Boyd, William Kenneth, ed., *Some Eighteenth Century Tracts Concerning North Carolina.* Raleigh: Edwards and Broughton Company, 1927.

[Caldwell, Joseph], *The Numbers of Carlton, Addressed to the People of North Carolina, on a Central Rail-Road Through the State.* New York: Published by G. Long, 1828.

Carman, Harry James, ed., *American Husbandry.* New York: Columbia University Press, 1939.

DeBow, James Dunwoody Brownson, *The Industrial Resources, Etc., of the Southern and Western States: . . . . ,* 3 vols. New York: D. Appleton and Company, 1854.

Easterby, James Harold, ed., *The South Carolina Rice Plantation: As Revealed in the Papers of Robert F. W. Allston.* Chicago: University of Chicago Press, 1945.

Edgar, Patrick Nisbett, *The American Race-Turf Register, Sportsman's Herald, and General Stud Book: . . . . ,* 2 vols. New York: Press of Henry Mason, 1833. In private possession, only first volume used.

Fries, Adelaide L., ed., *Records of the Moravians in North Carolina,* 6 vols. Raleigh: North Carolina Historical Commission, 1943.

Hamilton, Joseph Gregoire deRoulhac, ed., *The Papers of Thomas Ruffin,* 4 vols. Raleigh: Edwards & Broughton Printing Company, 1918.

Helper, Hinton Rowan, *The Impending Crisis of the South: How to Meet It.* New York: Burdick Brothers, 1857.

Holden, William Woods, *Address Delivered Before the Cumberland County Agricultural Society, at Fayetteville, November 3d., 1859.* Fayetteville: Printed by Edward J. Hale and Son, 1859.

Holmes, Francis Simmons, *The Southern Farmer and Market Gardener: Being a Compilation of Useful Articles on These Subjects, From the Most Approved Writers.* Charleston: Printed by Burges and James, 1842.

Hoyt, William Henry, ed., *The Papers of Archibald D. Murphey,* 2 vols. Raleigh: E. M. Uzzell & Company, 1914.

Hundley, Daniel Robinson, *Social Relations in Our Southern States.* New York: Henry B. Price, 1860.

[Jeffreys, George Washington], *Farmer's Own Book: A Series of Essays on Agriculture and Rural Affairs; in Forty-Seven Numbers.* By "Agricola." Raleigh: Printed and Published by Joseph Gales, 1819.

Kellar, Herbert Anthony, ed., *Solon Robinson: Pioneer and Agriculturist,* 2 vols. Indianapolis: Published by the Indiana Historical Bureau, 1936.

Lawson, John, *History of North Carolina.* Being a reprint of the copy in the North Carolina State Library, Raleigh, presented by President James Madison, in the year 1831. Charlotte: Observer Printing House, 1903.

Newsome, Albert Ray, ed., "Twelve North Carolina Counties in 1810-1811," *North Carolina Historical Review,* V (October, 1928), 413-446; VI (January, 1929), 67-99; (April, 1929), 171-189; (July, 1929), 281-309.

Phillips, Ulrich Bonnell, ed., *Plantation and Frontier Documents: 1649-1863.* Cleveland: The A. H. Clark Company, 1909.

*Proceedings of the Internal Improvement Convention, Which was Held at*

*Salisbury . . . 1836, with an Appendix.* Salisbury: Printed at the office of the *Watchman,* 1836.

Ruffin, Edmund, *Agricultural, Geological, and Descriptive Sketches of Lower North Carolina and the Similar Adjacent Lands.* Raleigh: Printed at the Institution for the Deaf, Dumb, and Blind, 1861.

Ruffin, Edmund, *Essays and Notes on Agriculture.* Richmond: J. W. Randolph, 1855.

*Transactions of the North-Carolina State Agricultural Society, for 1857.* Raleigh: Holden and Wilson, 1858.

Turner, J. A., *The Cotton Planters Manual: Being a Compilation of Facts From the Best Authorities on the Culture of Cotton; Its Natural History, Chemical Analysis, Trade and Consumption; and Embracing a History of Cotton and the Cotton Gin.* New York: C. M. Sexton and Company, 1857.

Wagstaff, Henry McGilbert, ed., *The Harrington Letters,* XIII, No. 2, *James Sprunt Historical Publications* (1914).

Wagstaff, Henry McGilbert, ed., *The Papers of John Steele,* 2 vols. Raleigh: Edwards & Broughton Printing Company, 1924.

D AGRICULTURAL AND OTHER PERIODICALS (Files in University of North Carolina Library.)

*American Farmer,* 21 vols. Baltimore and Washington, 1819-1839. Name changed to *Farmer and Gardener* in 1834. Name changed in 1839 to *American Farmer.* Only scattered issues available after 1839.

*Arator,* 3 vols. Raleigh, April, 1855-August, 1857.

*Carolina Cultivator,* 3 vols. Raleigh, March, 1855-October, 1857.

*Carolina Planter,* 1 vol. Columbia, South Carolina, January, 1840-January, 1841. This publication merged with the *Farmer's Register* in 1841.

*DeBow's Review,* 43 vols. New Orleans, Washington, Charleston, and Columbia, 1846-1884.

*Edgecombe Farm Journal,* 1 vol. Tarboro, September, 1860-March, 1861.

*Farmer's Advocate and Miscellaneous Reporter,* 4 vols. Jamestown, North Carolina, August, 1838-June, 1842. Scattered issues.

*Farmer's Journal,* 3 vols. Bath and Raleigh, April, 1852-December, 1854. Publication moved to Raleigh in August, 1853.

*Farmer's Register,* 11 vols. Shellbanks and Petersburg, 1833-1843.

*Niles' Weekly Register,* 76 vols. Baltimore and Philadelphia, 1811-1849. Title changed slightly several times.

*North Carolina Farmer,* 5 vols. Raleigh, June, 1845-October, 1849. Scattered issues.

*North Carolina Planter,* 4 vols. Raleigh, January, 1858-May, 1861.

*Southern Agriculturist,* 19 vols. Charleston, 1828-1846. Name changed in 1840 to *Southern Cabinet of Agriculture. . . .*

*Southern Cultivator,* 93 vols. Augusta, Athens, and Atlanta, 1843-1935. Only volumes to 1861 used.

E NEWSPAPERS (Files in University of North Carolina Library.)

Charlotte *Catawba Journal,* 1824-? Publication moved to Salisbury in 1828 and name changed to *Yadkin and Catawba Journal.* Scattered issues.

Edenton *Gazette,* 1800-1827. Name changed to *Edenton Gazette and North Carolina General Advertiser* and to *Edenton Gazette.* Scattered issues.

Edenton *Post-Angel or Universal Entertainment,* 1800-? Scattered issues.

Fayetteville *Gazette,* 1821-1822.

Fayetteville *North Carolina Minerva, and Fayetteville Advertiser,* 1796-1820. Publication moved to Raleigh in 1799 and name changed to *North Carolina Minerva and Raleigh Advertiser.* Name later changed to *North Carolina Minerva* and to *Minerva.* Photostatic copies. Scattered issues.

Greensboro *Patriot,* 1824-1909. Name changed several times. Only issues to 1861 used. Scattered issues.

Halifax *North Carolina Journal,* 1796-? Scattered issues.

Hillsboro *Recorder,* 1820-1896. Only issues to 1861 used. Scattered issues.

Newbern *North-Carolina Magazine; or Universal Intelligencer,* 1764-1765. Photostatic copies.

Newbern *State Gazette of North Carolina,* 1787-? Moved to Edenton in 1788. Photostatic copies. Scattered issues.

Newbern *North Carolina Gazette,* 1751-? Scattered issues.

Raleigh *Star,* 1809-1856. Name changed to *The Star and North Carolina State Gazette* and back to *Star.* Scattered issues.

Raleigh *North Carolina Standard,* 1835-1870. Name changed several times. Files to 1861 used only.

Salisbury *Carolina Watchman,* 1832-1890. Used to 1861 only. Scattered issues.

Salisbury *Western Carolinian,* 1820-1842.

Tarboro *Free Press,* 1824-1851. Name changed to *Press.*

Tarboro *Southerner,* 1852-1859.

Wilmington *Cape Fear Mercury,* 1769-1775. Photostatic copies. Scattered issues.

Wilmington *Hall's Wilmington Gazette,* 1797-1799. Photostatic copies.

Wilmington *Chronicle: and North Carolina Weekly Advertiser,* 1840-1849. Photostatic copies. Scattered issues.

F AUTOBIOGRAPHIES, DIARIES, JOURNALS, AND TRAVEL ACCOUNTS

Andrews, Evangeline Walker and Charles McLean Andrews, eds., *Journal of A Lady of Quality; Being the Narrative of a Journey from Scotland to the West Indies, North Carolina, and Portugal, in the years 1774 to 1776.* New Haven: Yale University Press, 1923. London: Humphrey Milford, Oxford University Press.

Asbury, Francis, *The Journal of the Reverend Francis Asbury, Bishop of the Methodist Episcopal Church, From August 7, 1771 to December 7, 1815,* 3 vols. New York: N. Bangs and T. Mason, 1821.

Bassett, John Spencer, ed., "General Slade's Journal of a Trip to Tennessee," *Historical Papers* published by the Trinity College Historical Society, Series VI (1906), 37-56.

Buckingham, James Silk, *The Slave States of America,* 2 vols. London and Paris: Fisher, Son & Company, 1842.

Featherstonhaugh, George William, *Excursion Through the Slave States, From Washington on the Potomac to the Frontier of Mexico; with*

*Sketches of Popular Manners and Geological Notices.* New York: Harper & Brothers, 1844.

Hall, Basil, *Travels in North America, in the Years 1827 and 1828,* 2 vols. Philadelphia: Carey, Lea & Carey, 1829.

Hall, Francis, *Travels in Canada and the United States, in 1816 and 1817.* Boston: Republished from the London edition by Wells and Lilly, 1818.

Lyell, Charles, *A Second Visit to the United States of America,* 2 vols. New York: Harper & Brothers, Publishers, London: John Murray, 1850.

Martineau, Harriett, *Society in America,* 3 vols., 2nd. ed. London: Saunders and Otley, 1837.

Nevins, Allen, ed., *American Social History as Recorded by British Travellers.* New York: Henry Holt and Company, 1923.

Olmsted, Frederick Law, *A Journey in the Back Country.* New York: Mason Brothers, 1860.

Olmsted, Frederick Law, *A Journey in the Seaboard Slave States with Remarks on Their Economy.* New York: Dix & Edwards, London: Sampson Low, Son & Company, 1856.

Parker, Robert J., "A Yankee in North Carolina: Observations of Thomas Oliver Larkin, 1821-1826," *North Carolina Historical Review,* XIV (October, 1937), 325-342.

Rowland, Buford, ed., "Some Observations on the Low Parts of North Carolina," *North Carolina Historical Review,* XV (April, 1938), 156-158.

Salley, Alexander Samuel, Junior, ed., *Narratives of Early Carolina, 1650-1708.* New York: Charles Scribner's Sons, 1911.

Schoepf, Johann David, *Travels in the Confederation, 1783-1784,* 2 vols. Translated from the German and edited by Alfred J. Morrison. Philadelphia: William J. Campbell, 1911.

Thwaites, Reuben Gold, ed., *Travels West of the Alleganies Made in 1793-1796 by Andre Michaux, in 1802 by F. A. Michaux; and in 1803 by Thaddeus Mason Harris, M.A.* Cleveland: The Arthur H. Clark Company, 1904.

Watson, Winslow Cossoul, ed., *Men and Times of the Revolution; or, Memoirs of Elkanah Watson, Including Journals of Travels in Europe and America, From 1777 to 1842, with his Correspondence with Public Men and Reminiscenses and Incidents of the Revolution.* New York: Dana and Company, Publishers, 1856.

## II SECONDARY SOURCES

### A STATE AND LOCAL HISTORIES

Allen, William Cicero, *History of Halifax County.* Boston: The Cornhill Company, 1918.

Arthur, John Preston, *Western North Carolina: A History.* Raleigh: Edwards and Broughton Printing Company, 1914.

Boyd, William Kenneth, *The Story of Durham: City of the New South.* Durham: Duke University Press, 1925.

Chandler, Julian Alvin Carroll, and others, eds., *The South in the Building of the Nation: A History of the Southern States Designed to Record the*

*South's Part in the Making of the American Nation. . . .*, 13 vols. Richmond: The Southern Historical Publication Society, 1909-1913.

Connor, Robert Digges Wimberly, *North Carolina: Rebuilding an Ancient Commonwealth,* 4 vols. Chicago and New York: The American Historical Society, Inc., 1929.

Fiske, John, *Old Virginia and Her Neighbors,* 2 vols. Boston and New York: Houghton, Mifflin and Company, 1900.

Hawks, Francis Lister, *History of North Carolina,* 2 vols. Fayetteville: Published by E. J. Hale & Son, 1859.

Henderson, Archibald, author and ed., *North Carolina: The Old North State and the New,* 5 vols. Chicago: The Lewis Publishing Company, 1941.

Lefler, Hugh Talmage, *A Guide to the Study and Reading of North Carolina History.* Chapel Hill: The University of North Carolina Press, 1955.

Lefler, Hugh Talmage and Albert Ray Newsome, *North Carolina: The History of a Southern State.* Chapel Hill: The University of North Carolina Press, 1954.

Leonard, Jacob Calvin, *Centennial History of Davidson County North Carolina.* Raleigh: Edwards and Broughton Company, 1927.

Sondley, Foster Alexander, *Asheville and Buncombe County.* Asheville: The Citizen Company, 1922.

Turner, Joseph Kelly and John Luther Bridgers, Junior, *History of Edgecombe County North Carolina.* Raleigh: Edwards and Broughton Printing Company, 1920.

B MONOGRAPHS AND SPECIAL STUDIES

Abernethy, Thomas Perkins, *From Frontier to Plantation in Tennessee: A Study in Frontier Democracy.* Chapel Hill: University of North Carolina Press, 1932.

Bassett, John Spencer, *Slavery and Servitude in the Colony of North Carolina,* Fourteenth Series, IV-V, *Johns Hopkins University Studies in Historical and Political Science.* Baltimore: The Johns Hopkins Press, 1896.

Bassett, John Spencer, *Slavery in the State of North Carolina,* Series XVII, Nos. 7-8, *Johns Hopkins University Studies in Historical and Political Science.* Baltimore: The Johns Hopkins Press, 1899.

Battle, Kemp Plummer, *History of the University of North Carolina,* 2 vols. Raleigh: Edwards and Broughton Printing Company, 1907.

Beer, George Louis, *The Old Colonial System, 1660-1754,* 2 vols. New York: The Macmillan Company, 1912.

Bennett, Hugh Hammond, *The Soils and Agriculture of the Southern States.* New York: The Macmillan Company, 1921.

Bidwell, Percy Wells and John I. Falconer, *History of Agriculture in the Northern United States, 1620-1860.* Washington: Published by the Carnegie Institution of Washington, 1925.

Bogart, Ernest Ludlow, *Economic History of the American People,* 2nd. ed. New York: Longmans, Green and Company, 1935.

Bond, Beverly Waugh, *The Quit-Rent System in the American Colonies.* New Haven: Yale University Press, 1919.

Brooks, Eugene Clyde, *The Story of Cotton and the Development of the Cotton States.* New York: Rand, McNally and Company, 1911.

Brown, Cecil Kenneth, *A State Movement in Railroad Development: The Story of North Carolina's First Effort to Establish an East and West Trunk Line Railroad.* Chapel Hill: University of North Carolina Press, 1928.

Brown, Cecil Kenneth, *The State Highway System of North Carolina, Its Evolution and Present Status.* Chapel Hill: University of North Carolina Press, 1931.

Burlingame, Roger, *March of Iron Men: A Social History of Union Through Invention.* New York: Charles Scribner's Sons, 1938.

Carrier, Lyman, *The Beginnings of Agriculture in America,* 1st. ed. New York: McGraw-Hill Book Company, Inc., 1923.

Clewell, John Henry, *History of Wachovia in North Carolina. The Unitas Fratum or Moravian Church in North Carolina During a Century and a Half, 1752-1902, From the Original German and English Manuscripts and Records in the Wachovia Archives, Salem, North Carolina.* New York: Doubleday, Page and Company, 1902.

Combs, Joseph Franklin, *Growing Pastures in the South.* Chapel Hill: University of North Carolina Press, 1936.

Connor, Robert Digges Wimberly, "Race Elements in the White Population of North Carolina," *North Carolina State Normal & Industrial College Historical Publications,* No. 1, 1920.

Cool, R. C., *The Scuppernong Grape: Its Growth and Care Under Vineyard Conditions.* Raleigh: Edwards and Broughton Printing Company, 1913.

Craven, Avery Odell, *Edmund Ruffin, Southerner: A Study in Secession.* New York and London: D. Appleton and Company, 1932.

Craven, Avery Odell, *Soil Exhaustion as a Factor in the Agricultural History of Virginia and Maryland, 1606-1860.* Urbana: University of Illinois, 1926.

Crittenden, Charles Christopher, *The Commerce of North Carolina, 1763-1789.* New Haven: Yale University Press, 1936.

Demaree, Albert Lowther, *The American Agricultural Press 1819-1860.* New York: Columbia University Press, 1941.

Doar, David, *Rice and Rice Planting in the South Carolina Low Country.* Charleston: Published by the Charleston Museum, 1936.

Dodd, William Edward, *Expansion and Conflict.* New York: Houghton Mifflin Company, 1915.

Dodd, William Edward, *The Cotton Kingdom: A Chronicle of the Old South.* New Haven: Yale University Press, 1919.

Duggar, Benjamin M., "Grass and Forage Crop Farming in the South," *The South in the Building of the Nation,* V, 222-229.

Duggar, John Frederick, *Southern Forage Crops.* New York: The Macmillan Company, 1925.

Edwards, Everett Eugene, *A Bibliography of the History of Agriculture in the United States.* United States Department of Agriculture, Miscellaneous Publication No. 84. Washington: Government Printing Office, 1930.

Emory, Samuel Thomas, *Bright Tobacco in the Agriculture, Industry and Foreign Trade of North Carolina.* Chicago: University of Chicago, 1939.

Fleming, Walter Lynwood, "The Slave Labor System in the Ante-Bellum South," *The South in the Building of the Nation,* V, 104-120.

Franklin, John Hope, *The Free Negro in North Carolina, 1790-1860.* Chapel Hill: University of North Carolina Press, 1943.

Gray, Lewis Cecil, *History of Agriculture in the Southern United States to 1860,* 2 vols. Washington: Published by the Carnegie Institution of Washington, 1933.

Hamilton, Joseph Gregoire deRoulhac, *Party Politics in North Carolina 1835-1860, James Sprunt Historical Publications,* XV, Nos. 1 and 2, 1916.

Hammond, Matthew Brown, "The History of Cotton Planting in the South," *The South in the Building of the Nation,* V, 197-212.

Hawk, Emory Quinter, *Economic History of the South.* New York: Prentice-Hall, Inc., 1934.

Heyward, Duncan Clinch, *Seed From Madagascar.* Chapel Hill: University of North Carolina Press, 1937.

Hobbs, Samuel Huntington, Junior, *North Carolina: Economic and Social.* Chapel Hill: University of North Carolina Press, 1930.

Holder, Brantson Beeson, "The Three Banks of the State of North Carolina, 1810-1872." Unpublished doctoral dissertation, University of North Carolina, 1937.

Hulbert, Archer Butler, *Soil; Its Influence on the History of the United States: With Special Reference to Migration and the Scientific Study of Local History.* New Haven: Yale University Press, 1930.

Hulbert, Archer Butler, *The Paths of Inland Commerce: A Chronicle of Trail, Road, and Waterway.* New Haven: Yale University Press, 1921.

Hunt, Thomas F., "The History of Cereal Farming in the South," *The South in the Building of the Nation,* V, 212-222.

Jacobstein, Meyer, *The Tobacco Industry in the United States, Columbia University Studies in History, Economics, and Public Law,* XXVI, No. 3, 1907.

Jacobstein, Meyer, "Tobacco Culture in the South," *The South in the Building of the Nation,* V, 158-169.

Johnson, Guion Griffis, *Ante-Bellum North Carolina: A Social History.* Chapel Hill: University of North Carolina Press, 1933.

Kennedy, Edwin Wexler, *Quit-Rents and Currency in North Carolina, 1663-1776.* Baltimore: The J. W. Bond Company, 1902.

Killebrew, Joseph Buckner and Herbert Myrick, *Tobacco Leaf: Its Culture and Cure, Marketing and Manufacture.* New York: Orange Judd Company, 1897.

Konkle, Burton Alva, *John Motley Morehead and the Development of North Carolina, 1796-1866.* Philadelphia: William J. Campbell, 1922.

Mell, Patrick H., "Rice Planting in the Agricultural Development of the South," *The South in the Building of the Nation,* V, 169-177.

Meyer, Balthasar Henry, ed., with associates, *History of Transportation in the United States Before 1860*. Washington: Published by the Carnegie Institution of Washington, 1917.

Mirsky, Jeannette and Allan Nevins, *The World of Eli Whitney*. New York: The Macmillan Company, 1952.

Moore, Thomas Brent, "Southern Hemp and Flax Production," *The South in the Building of the Nation,* V, 229-235.

Neely, Wayne Caldwell, *The Agricultural Fair*. New York: Columbia University Press, 1935.

Nettels, Curtis Potts, *The Roots of American Civilization*. New York: F. S. Crofts & Company, 1938.

Nevins, Allan, *The American States During and After the Revolution, 1775-1789*. New York: The Macmillan Company, 1924.

Nixon, Joseph R., *The German Settlers in Lincoln County and Western North Carolina, James Sprunt Historical Publications,* XI, No. 2, 1912.

Phillips, Ulrich Bonnell, *American Negro Slavery: A Survey of the Supply, Employment and Control of Negro Labor as Determined by the Plantation Regime*. New York: D. Appleton and Company, 1918.

Phillips, Ulrich Bonnell, *A History of Transportation in the Eastern Cotton Belt to 1860*. New York: Columbia University Press, 1908.

Phillips, Ulrich Bonnell, "Financial Crisis in the Ante-Bellum South," *The South in the Building of the Nation,* V, 435-441.

Phillips, Ulrich Bonnell, *Life and Labor in the Old South*. Boston: Little, Brown, and Company, 1929.

Robert, Joseph Clarke, *The Tobacco Kingdom: Plantation, Market, and Factory in Virginia and North Carolina, 1800-1860*. Durham: Duke University Press, 1938.

Rommel, George M., "The Animal Industry of the South," *The South in the Building of the Nation,* V, 242-257.

Sanford, Albert Hart, *The Story of Agriculture in the United States*. Boston, New York, Chicago: D. C. Heath and Company, Publishers, 1916.

Scherer, James Augustin Brown, *Cotton as a World Power*. New York: Frederick A. Stokes, 1916.

Schmidt, Louis Bernard and Earle Dudley Ross, eds., *Readings in the Economic History of American Agriculture*. New York: The Macmillan Company, 1925.

Smith, Abbot Emerson, *Colonists in Bondage: White Servitude and Convict Labor in America, 1607-1776*. Chapel Hill: University of North Carolina Press, 1947.

Soule, Andrew M., "Vegetables, Fruit and Nursery Products and Truck Farming in the South," *The South in the Building of the Nation,* V, 236-242.

Sprunt, James, *Chronicles of the Cape Fear River, 1660-1916,* 2nd. ed. Raleigh: Edwards and Broughton Printing Company, 1916.

Stone, Alfred Holt, "The Influence of the Factorage System, Foreign and Domestic, on Southern Agriculture," *The South in the Building of the Nation,* V, 398-404.

Taylor, Rosser Howard, *Slaveholding in North Carolina: An Economic View, James Sprunt Historical Publications,* XVIII, Nos. 1 and 2, 1926.

Thompson, James Westfall, *A History of Livestock Raising in the United States, 1607-1860.* Agricultural History Series No. 5. Publication of the United States Department of Agriculture, 1942.

True, Alfred Charles, *A History of Agricultural Experimentation and Research in the United States, 1607-1925, Including a History of the United States Department of Agriculture.* Miscellaneous Publication No. 251, United States Department of Agriculture. Washington: Government Printing Office, 1937.

Turlington, Sarah Woodall, "Steam Navigation in North Carolina Prior to 1860." Unpublished masters thesis, University of North Carolina, 1933.

Uzzell, William Edward, "History of Tobacco in North Carolina." Unpublished masters thesis, University of North Carolina, 1929.

Vance, Rupert Bayless, *Human Geography of the South: A Study in Regional Resources and Human Adequacy.* Chapel Hill: University of North Carolina Press, 1932.

Voorhees, Edward Burnett, *Forage Crops For Soiling, Silage, Hay, and Pasture,* 2nd. ed. New York: The Macmillan Company, 1911.

Wall, Bennett Harrison, "Ebenezer Pettigrew, An Economic Study of an Ante-Bellum Planter." Unpublished doctoral dissertation, University of North Carolina, 1947.

Weaver, Charles Clinton, *Internal Improvements in North Carolina Previous to 1860. Johns Hopkins University Studies in History and Political Science,* Series XXI, Nos. 3-4. Baltimore: Johns Hopkins Press, 1903.

Wertenbaker, Thomas Jefferson, *The Planters of Colonial Virginia.* Princeton: Princeton University Press, 1922.

Whitener, Daniel Jay, *Prohibition in North Carolina, 1715-1945, James Sprunt Studies in History and Political Science,* XXVII, 1945.

Whitney, Milton, "Climatology and Soils," *House Document No. 267,* 54th Congress (1896-1897).

Whitney, Milton, *Soil and Civilization: A Modern Concept of the Soil and the Historical Development of Agriculture.* New York: D. Van Nostrand Company, Inc., 1925.

Wilkin, Helen D., "The Promotion of Agriculture in North Carolina, 1810-1860." Unpublished masters thesis, University of North Carolina, 1941.

Williams, James William, "Emigration From North Carolina: 1789-1860." Unpublished masters thesis, University of North Carolina, 1939.

Winkler, John Kennedy, *Tobacco Tycoon: The Story of James Buchanan Duke.* New York: Random House, 1942.

Young, Wade Phillips, "A History of Agricultural Education in North Carolina." Unpublished doctoral dissertation, University of North Carolina, 1934.

C PERIODICAL ARTICLES

Anderson, Russell H., "Grain Drills Through Thirty-Nine Centuries," *Agricultural History,* X (October, 1936), 157-205.

Bassett, John Spencer, "Landholding in Colonial North Carolina," *The Law Quarterly Review,* XI (April, 1895), 154-166.

Bassett, John Spencer, "The Influence of Coast Line and Rivers on North

Carolina," *Annual Report of the American Historical Association,* I (1908), 58-61.

Bonner, James Calvin, "The Plantation Overseer and Southern Nationalism As Revealed in the Career of Garland D. Harmon," *Agricultural History,* XIX (January, 1945), 1-11.

Boyd, William Kenneth, "Currency and Banking in North Carolina, 1790-1836," *Historical Papers* published by the Trinity College Historical Society, Series X, 1914, 52-86.

Carman, Harry James, "Jesse Buel, Early Nineteenth Century Agricultural Reformer," *Agricultural History,* XVII (January, 1943), 1-13.

Cathey, Cornelius Oliver, "Sidney Weller: Ante-Bellum Promoter of Agricultural Reform," *North Carolina Historical Review,* XXXI (January, 1954), 1-17.

Clonts, F. W., "Travel and Transportation in Colonial North Carolina," *North Carolina Historical Review,* III (January, 1926), 16-35.

Connor, Robert Digges Wimberly, "John Motley Morehead: Architect and Builder of Public Works," *North Carolina Booklet,* XII (January, 1913), 173-192.

Connor, Robert Digges Wimberly, "The Settlement of the Cape Fear," *The South Atlantic Quarterly,* VI (July, 1907), 272-287.

Craven, Avery Odell, "The Agricultural Reformers of the Ante-Bellum South," *American Historical Review,* XXXIII (January, 1928), 302-314.

Crittenden, Charles Christopher, "North Carolina Newspapers Before 1790," *The James Sprunt Historical Studies,* XX, No. 1 (1928), 7-83.

Franklin, W. Neil, "Agriculture in Colonial North Carolina," *North Carolina Historical Review,* III (October, 1926), 539-574.

Galloway, Beverly Thomas, "Plant Pathology; A Review of the Development of the Science in the United States," *Agricultural History,* II (April, 1928), 49-60.

Gehrke, William Herman, "The Ante-Bellum Agriculture of the Germans in North Carolina," *Agricultural History,* IX (July, 1935), 143-160.

Harrell, Isaac Samuel, "Gates County to 1860," *Historical Papers,* published by the Trinity College Historical Society, Series XII (1916), 57-106.

Leavitt, Charles T., "Attempts to Improve Cattle Breeds in the United States, 1790-1860," *Agricultural History,* VII (April, 1933), 51-67.

Lemert, Ben Franklin, "Geographic Influences in the History of North Carolina," *North Carolina Historical Review,* XII (October, 1935), 297-319.

Morgan, J. Allen, "State Aid to Transportation in North Carolina," *North Carolina Booklet,* X (January, 1911), 122-154.

Morris, Francis Grave and Phyllis Mary Morris, "Economic Conditions in North Carolina About 1780," *North Carolina Historical Review,* XVI (April, 1939), 107-133; (July, 1939), 296-327.

Russel, Robert R., "The Effects of Slavery Upon Nonslaveholders in the Ante Bellum South," *Agricultural History,* XV (April, 1941), 112-126.

Sitterson, Joseph Carlyle, "Economic Sectionalism in Ante-Bellum North Carolina," *North Carolina Historical Review,* XVI (April, 1939), 134-146.

Starling, Robert B., "The Plank Road Movement in North Carolina," *North Carolina Historical Review,* XVI (January, 1939), 1-22, (April, 1939), 147-173.

Tatter, Henry, "State and Federal Land Policy During the Confederation Period," *Agricultural History,* IX (October, 1935), 176-186.

Turner, James K., "Slavery in Edgecombe County," *Historical Papers,* published by the Trinity College Historical Society, Series XII (1916), 5-36.

# INDEX

www.ingramcontent.com/pod-product-compliance
Lightning Source LLC
LaVergne TN
LVHW090354210926
848480LV00002B/334